Martin J. Forrest
Recycling of Polyethylene Terephthalate

Also of interest

Rubber Analyis
Characterisation, Failure Diagnosis and Reverse Engineering
Forrest, 2019
ISBN 978-3-11-064027-4, e-ISBN 978-3-11-064028-1

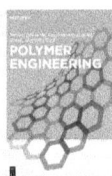

Polymer Engineering
Tylkowski, Wieszczycka, Jastrzab (Eds.), 2017
ISBN 978-3-11-046828-1, e-ISBN 978-3-11-046974-5

Green Chemistry and Technologies
Zhang, Gong, Bin (Eds.), 2018
ISBN 978-3-11-047861-7, e-ISBN 978-3-11-047931-7

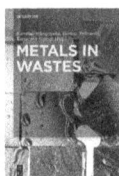

Metals in Wastes
Wieszczycka, Tylkowski, Staszak, 2018
ISBN 978-3-11-054628-6, e-ISBN 978-3-11-054706-1

e-Polymers
Editor-in-Chief: Seema Agarwal
ISSN 2197-4586
e-ISSN 1618-7229

Martin J. Forrest

Recycling of Polyethylene Terephthalate

2nd Edition

DE GRUYTER

Author
Dr. Martin J. Forrest
Shrewsbury Laboratory
Shrewsbury
SY4 4NR
Great Britian

ISBN 978-3-11-064029-8
e-ISBN (PDF) 978-3-11-064030-4
e-ISBN (EPUB) 978-3-11-064045-8

Library of Congress Control Number: 2018966614

Bibliographic information published by the Deutsche Nationalbibliothek
The Deutsche Nationalbibliothek lists this publication in the Deutsche Nationalbibliografie;
detailed bibliographic data are available on the Internet at http://dnb.dnb.de.

© 2019 Walter de Gruyter GmbH, Berlin/Boston
Typesetting: Integra Software Services Pvt. Ltd.
Printing and binding: CPI books GmbH, Leck
Cover image: stevanovicigor / iStock / Getty Images Plus

www.degruyter.com

To Mum and Dad, who made it all possible.

Preface

This book is a review of the current state of the art and covers the main areas of interest, research and commercial exploitation associated with the recycling and re-use of polyethylene terephthalate (PET). It places the situation that exists today into an historical perspective and describes the various factors (environmental, legislative, economic and social) that have helped drive innovation and research activity to the point where several potential routes for the recycling of PET on the supply side have been investigated and evaluated with respect to their technical and economic merits and their compliance with the applicable standards and regulations. The routes available include 'physical' processes concerned with the production of high-quality recycled polyethylene terephthalate (rPET) [e.g., high-quality washed flake (HQWF)] that can be used very successfully to produce end-products such as rPET fibres and strapping, and 'physical' and 'chemical' super-clean systems that can generate rPET capable of meeting the European Union (EU) and the US Food and Drug Administration (FDA) regulatory requirements to produce food contact products. The principal examples of these technologies are covered, along with the provision of information regarding the analysis and characterisation methods available to ensure that the rPET can meet any applicable regulatory and technological requirements that apply to the intended final product.

With the pressing need in certain areas of the world to re-use end- of-life PET products as a result of legislation, and the increased availability of high-quality rPET, there has been a considerable amount of research activity taking place on the consumption side. This book reviews the numerous routes that have been evaluated and, in several cases, commercially exploited for the use of rPET. These include routes that address the important food-packaging sector and which seek to recycle food-grade PET products into new food contact products, and the 'bottle-to-bottle' processes are a very successful example of these routes. Also covered are numerous other recycling options, from those that look to recover energy, or to generate reactive low-molecular weight intermediates (e.g., monomers) which can be used to create new materials (e.g., thermoset resins and coatings), to work that has taken place in the creation of rPET blends with other polymeric materials (e.g., rubbers and thermoplastics) or the production of fibres for the manufacture of clothing and other articles.

To set the technological progress and innovations that has been made into an economic context, an overview of the market for rPET is provided, and vital areas of the general recycling infrastructure, such as improving the sorting and separation of 'difficult' post-consumer PET products (e.g., black and laminated products) is addressed. Also, to help place the recycling of PET into a general recycling context, there is also a section that provides an update on the regulations in the EU and the US that apply to post-consumer plastics. To demonstrate how advances are being made to access the proportion of PET products in the recycling stream that are

https://doi.org/10.1515/9783110640304-201

inaccessible (e.g., black pigmented products), there is a section on the developments taking place to improve the sorting and separation of post-consumer plastics.

To write this book, the up-to-date, extensive information present in the Smithers Rapra Polymer Library has been used. This source has been complemented by other sources of information, for example, the published findings of relevant research projects, references from trade literature, and the considerable in-house expertise that has been gained at Smithers Rapra due to the research that the company has been involved with in this field. In addition to providing a summary of the technical aspects of all the areas that are addressed, because of the availability of the resources of the Smithers Rapra Polymer Library, this book also functions as a comprehensive, up-to-date literature review of the subject.

Acknowledgements

Many thanks to my colleagues at Smithers Information who have assisted me in the writing and production of this book, in particular Eleanor Garmson and Helene Chavaroche, who have provided advice and used their expertise with Arshad Makhdum to edit this book for publication.

Thanks also go to the Waste Resource Action Programme (WRAP) (http://www.wrap.org.uk/) who have kindly provided me with permission to include their photographs and images in the book.

https://doi.org/10.1515/9783110640304-202

Contents

1 Introduction to polyethylene terephthalate recycling

The recycling of polyethylene terephthalate (PET) has been carried out for many years, with Saint Jude Polymers reported as being the first company in the USA to set up a process to recycle PET bottles in 1976. In this process, the bottles were recycled into plastic strapping and paintbrush bristles, and 1 year later the company began to produce pelletised recycled polyethylene terephthalate (rPET) for the general market. Other companies, particularly Wellman Incorporated, started to recycle PET into other products (e.g., carpet fibres) and the rPET industry continued to expand throughout the 1980s and 1990s. Given its large-scale use for food-packaging products, it was only a matter of time before this market was targeted. By the 1990s, with the appropriate recycling technology now available, 'letters of no objection' started to be issued by the US Food and Drug Administration for the use of rPET in food contact packaging applications [1].

However, in common with other major manufacturing industries, such as the rubber industry, the pressure on the plastics industry and the many users of its products to recycle plastic has increased dramatically over the last 25 years or so due to a combination of economic, environmental, societal and legislative factors. The need to conserve natural resources, coupled with the publication of important legislation, such as the Landfill Directive (1999/31/EC) and the Packaging and Packaging Waste Directive (94/62/EC) in the European Union (EU), have stimulated the search for technologies and manufacturing processes that can recycle and re-use waste plastic. When it comes to recycling, the significant advantage that plastics, such as PET, have over thermoset materials, such as rubber and thermosetting resins (e.g., phenolics and epoxies) is that, once separated and decontaminated, they can be reprocessed using the same processing techniques as their virgin equivalents.

In the EU, the debate on how to continue to increase the amount of plastic that is recycled and re-used for the benefit of the community is continuing. A demonstration of the intention to increase the focus on plastic recycling was provided by the recent green paper by the European Commission (EC) indicating that it planned to revise legislation affecting this sector [2]. The main focus of this legislation is likely to be greater emphasis on recycling targets, solving the problem of the landfilling of plastics and increasing the quality of recyclates. Some of the proposals in this green paper were:

- Phasing out landfilling by 2025 for recyclable materials (e.g., plastics, metals, paper, and bio-waste).
- Recycling and preparing for re-use of packaging waste to be increased to 80% by 2030, with material-specific targets set to increase gradually between 2020 and 2030, to reach:

https://doi.org/10.1515/9783110640304-001

- 60% for plastics by the end of 2030.
- 80% for wood and 90% ferrous metal, aluminium and glass by the end of 2030.
- 90% for paper by the end of 2025.
- Recycling and preparing for re-use of municipal waste to be increased to 70% by 2030.

It has been reported [3] that the comments that were received by the EC after the publication of this green paper showed that there was strong support for promoting monomaterials and improving the design of plastics to increase recyclability. In addition, the responses obtained showed that there were split views about biodegradable and bio-based plastics, that more consumer information on plastics recycling was needed, and that better waste collection and sorting infrastructure was required.

Due to these considerations, increased funding has become available from several sources, including national and regional governments, and the area of plastics recycling is now an extremely active one with areas such as improved sorting and identification systems and the production of food-grade recyclate being of particular interest to researchers. It is possible to divide different types of recycling activities and processes for waste plastics into four broad generic categories:

- *Primary*: Reprocessing into materials and products having properties that are the same (or at least comparable) to the original material or product.
- *Secondary*: Where the recycled plastic is made into products that do not have (or need to have) properties that are the same or comparable to the original product.
- *Tertiary*: The reduction of the recycled plastic into small chemical units (i.e., molecules) that can then be recycled into new materials and products by routes such as re-polymerisation.
- *Quaternary*: The recovery of the energy inherently present in recycled plastic by methods such as incineration or the burning of fuel products that are derived from pyrolysis processes.

As this book demonstrates, all four of the recycling routes described above have been explored in the search for new and effective ways to recycle PET. The relative ease with which PET can be depolymerised by hydrolysis has resulted in variants of the Tertiary route often being used as part of the purification process in order to produce food grade rPET. Evidently, in this case, because the purpose is to re-create high-quality, high-molecular weight rPET [e.g., for bottle-to-bottle (B2B) recycling], this particular recycling stream could also be regarded as a form of the *Primary* route.

PET is an important class of plastic material and it is used for several applications (Chapters 8 and 9), including:

- Food packaging (e.g., bottles, trays and film)
- Non-food packaging (e.g., containers for cosmetics, healthcare, and detergents)

- Strapping products
- Fibres (e.g., for clothing and bags)
- Non-woven fabrics
- Carpets

Of the end-applications for PET, packaging, both food [4] (Figure 1.1) and non-food, is the most important. It is in this area that most of the recycling effort has been concentrated, particularly in the food-packaging sector, where several new technologies (often referred to as 'super-clean' recycling processes) have been developed to enable post-consumer waste (particularly PET bottles) to be re-recycled back into food grade pellets and products (Chapter 6). The market in Europe and the rest of the world for PET B2B recycling is growing rapidly due to several factors, including government initiatives intended to facilitate the meeting of recycling targets, and technological advances which have seen the development of several chemical and physical processes that can regenerate food-grade PET. In Europe, there are many super-clean recycling processes (mostly for PET) that are awaiting consideration by the European Food Safety Authority (EFSA). This fact has not stopped them being used commercially as the dossiers to show that they meet the EU Plastics Recycling for Food Use Regulation 282/2008/EC were compiled for these submissions, and EFSA permits the processes to be used while the assessment is being carried out. EFSA adopted its first three scientific opinions on the safety of processes to recycle PET for use in food contact materials in 2012 [5]. These processes were considered by EFSA not to give rise to safety concerns if operated under well-defined and controlled conditions as outlined in 282/2008/EC. These three 'opinions', which were to be the first in a series to be published by EFSA (Chapter 6), covered ten recycling processes, which were grouped according to the particular type of recycling technology that they employed, as shown below:

- Four recycling processes based on VACUREMA Prime® technology.
- Five recycling processes based on Starlinger IV+® technology.
- Recycling process 'PETUK SSP'.

The relative importance of PET in the packaging market is shown in data provided by Mergers Alliance [6]. They have described the overall market structure for plastic packaging in the EU (Table 1.1) and, of the ≈18 million tonnes of plastic that are converted into all types of packaging (i.e., food and non-food), PET has an overall share of 8.6%.

The average rate of recycling for plastic packaging in the EU is 26% and, as described in Chapters 2 and 3, plastic bottles are the leading source of plastic for recycling, accounting for more than half of all plastic recycled in 2012, and the biggest growth opportunities for recycled plastics in packaging will come from PET. This scenario was illustrated by Mergers Alliance [6] who estimated in 2012 that, at the

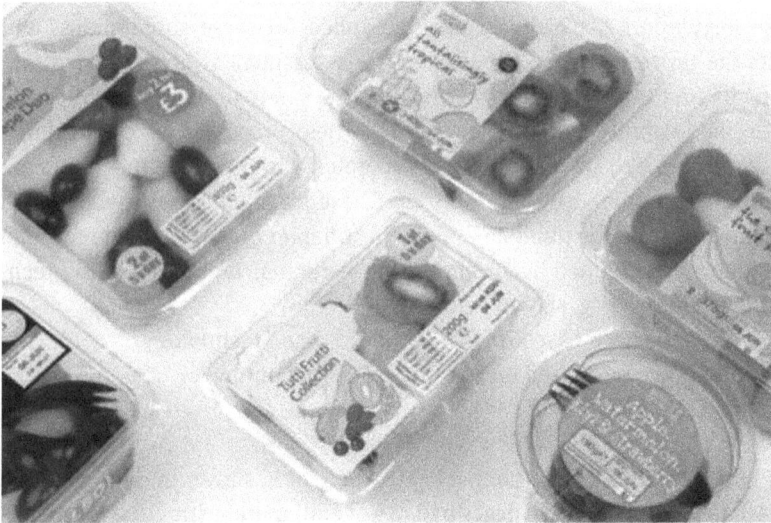

Figure 1.1: Typical PET food tray packaging in the UK. Reproduced with permission from the Waste and Resources Action Programme (WRAP), Banbury, UK. ©WRAP.

end of its life, 66% of the ≈18 million tonnes of plastic packaging was recycled (into new products or energy recovery), but that 34% was sent to landfill.

Industry is assisting the growth of rPET in food packaging by developing new equipment that can tackle production issues. For example, it was reported in

Table 1.1: Types of plastic used in the packaging industry within the EU [6].

Plastic type	Share of total packaging (%)
PS	4.7
Polyurethane	6.7
PET	8.6
PVC	11.3
Low-density PE	11.5
HDPE	17.9
PP	18.6
Other	20.7

PE: Polyethylene
PP: Polypropylene
PS: Polystyrene
PVC: Polyvinyl chloride
Reproduced with permission from Mergers Alliance, Plastics Europe MRG, Rexam, 2012. ©2012, Mergers Alliance [6]
http://www.mergers-alliance.com

PETplanet Insider [7] that Sipa had developed injection-moulding equipment having a range of cavities from 96 down to 24 to produce PET preforms that contained high levels of post-consumer PET flake. The Sipa Preform production systems can produce preforms that contain levels of rPET ≤100%. The article, which was published in 2011, also mentioned that it was possible to use ≤50% rPET in bottles for mineral water sold in the EU, but that several companies in North America were already producing mineral water bottles from 100% rPET.

Baker [8] has stated the benefits that plastics bring to packaging materials, including the reduction in weight (which reduces the 'carbon footprint' of transportation) and the functionality that enables re-closable packs to be produced (which helps reduce food waste to improve sustainability). The recyclability of plastics is also covered, with mention of those that are recycled widely [PET and high-density polyethylene (HDPE) milk bottles] and those that can be recycled easily [non-milk HDPE and PP]. The launch of Akzo Nobels, Dulux, first-to-market 2.5-l and 5-l PP paint cans in 2012, that are produced using 25% recycled PP, are cited as an example of the progress that is being made in finding new applications for recycled plastics. Producing packaging that has better barrier properties can also reduce food waste by helping retailers keep products fresher for longer. Holfeld Plastics Ltd, in a press release in 2013 [9], announced that they had released a new range (the 'plusS range') of sustainable rPET/recycled PE food trays that could be used to package poultry and fish products.

On a similar theme, Deschamps has reported on the continuing trend away from traditional packaging materials such as glass, metal and paperboard cartons to rigid plastics due to their advantages, such as being shatterproof and having a lighter weight [10]. The article also refers to the trend towards recyclability, with PET, a plastic with a relatively well-developed infrastructure (Chapters 3, 5 and 6), overtaking PE as the leading plastic for rigid packaging, and that Smithers PIRA have reported that suppliers across all packaging sectors are looking to add value through innovation by integrating lightweighting, recycling, biodegradability and sustainability. Examples of these products are cited, including a 300-ml PET bottle for Marks & Spencer's Essential Extracts personal care range that contains 30% rPET and an innovative PET jar (Big Mouth) produced by Amcor for hot filled foods which is 86% lighter than glass and 34% lighter than traditional heat-set PET. An example of the move away from traditional materials to plastics is outlined in an article in the *International Bottler & Packager* journal [11]. The article highlights the fact that breweries are making greater efforts to use new lightweight plastic containers as low-cost, environmentally friendly alternatives to their stocks of traditional metal kegs. The company Petainer is said to be leading the way in this transition by manufacturing an innovative, economical and fully recyclable PET container that is designed for one-way use to fit in with traditional industry working practices. The container is also supplied with low-cost fittings that allow it to be connected to existing tapping systems for draught beer. The product is claimed to enable breweries to cut costs and reduce environmental impacts in areas such as transport, use of water and energy, and waste generation. New recyclable PET products are also being launched to package wine. Vinoware is a totally recyclable 100% virgin PET container comprising practical and easily stored four-packs of stem-less wine glasses. A foil seal preserves the freshness of the wine in each glass and the four glasses together hold the equivalent of a full 750-ml bottle [12].

In the UK, the rate of plastic recycling in 2012 was 32%, and the government announced that it wanted this to be increased to 57% over the next 5 years. Such governmental announcements are not always well received by industry, and the British Plastics Federation were reported at the time as saying that this new target represented an additional tax on industry of £70 million over that period [13]. The industry regarded the target to be achievable only if adequate recycling facilities were available and accessible, something that was not thought to be the case in the UK at the time, because the infrastructure had not been developed to manage the separation of streams of mixed plastic waste economically. This meant that the target was likely to be met only in the short term by exporting recovered plastic for recycling. It is for these reasons that, over recent years, the UK recycling agency WRAP has been supporting higher rates of recycling of plastic packaging by use of a Mixed Plastics Loan Fund [14]. In this way, WRAP aims to increase the UK's recycling capacity for packaging of rigid plastics by 100,000 tonnes per year without prejudicing quality and achieving optimal economic and environmental outcomes. This improvement in infrastructure has included non-bottle rigid plastic packaging, such as PET food trays. Households in the UK produce ≈1.7 million tonnes of plastic packaging waste each year and, of this value, more than

500,000 tonnes are bottles, a significant proportion of which will be PET bottles [15, 16]. In terms of value, PET and HDPE are the common and profitable plastics to recycle because these are relatively easy to sell and generate good returns (Chapter 2).

An example of how the infrastructure in countries such as the UK has been improving over recent years has been the opening of new sorting and recycling facilities. In April 2011, Biffa Waste Group opened in Redcar in the UK what it said at the time was Europe's first integrated sorting and recycling facility for mixed plastics packaging [17]. The plant, which was opened with the aid of a grant from WRAP of £1.187 million, took mixed waste plastics such as pots, tubs, trays and bottles, washed them and then sorted them into the following streams:
–   PP Natural
–   PP Jazz
–   PE Natural
–   PE Jazz
–   Black PE and PP mix
–   PET Jazz
–   High-impact PS sheet Jazz
    where Jazz = mixed colour.

In the same article, Biffa were reported to be aiming to operate at the full capacity of 20,000 tonnes per year by April 2012. The company expected a lot of its recycled plastic to be used in the packaging of fast-moving consumer goods such as hair products, paint products, and products for other markets (e.g., horticulture). The plant was re-opened in late 2013 after having undergone a major upgrade [18]. This Biffa facility followed on from the opening by J&A Young in August 2008 of the first recycling facility in the UK capable of sorting mixed plastics on a large scale. The plant was reported as having a capacity of 78,000 tonnes per year and was able to process post-consumer plastic in bottle and non-bottle forms [19].

One area where there are still advances to be made in the UK (and in many other countries) is in the collection and recycling of plastic films. PET film is often found in domestic waste because it is used for high-temperature food applications (e.g., covering oven-ready meals). Plastic film made a major contribution to the 1 million tonnes of plastic waste land-filled in the UK in 2011, with the others being black plastic (due to sorting problems) (Chapter 5), and the lack of high-value markets for non-bottle plastic. As mentioned above, WRAP in the UK is funding research and development (R&D) work to try and address these issues [20].

With regard to the US market, according to the US Environmental Protection Agency [21] the overall combined recovery rate in the US for glass, metal, paper, board and plastic in 2011 was 48.3%, with the recovery rate for plastics being only ≈12%. The overall recycling rate for these materials was 24.0% (Chapter 2). A review was published in 2013 by the American Chemistry Council [22] of the progress made in the USA over the last 20 years with regard to the recycling of post-consumer plastics. The data were presented in the form of the categories of materials involved, and included soft drinks and other bottles made of PET, rigid products (e.g., containers) produced from HDPE, PP, PET, PS and PVC, and other packaging materials (e.g., films from PE). The review also covered use of recycled plastics for the manufacture of new products by various manufacturing sectors (e.g., automotive, clothing and electronics).

Recycling of plastics lags behind other global recycling markets (particularly for metal and glass) but has improved its status in recent years due to regulatory and environmental pressures and influences, and technological advances (e.g., improved collection regimens, sorting techniques and recycling processes). Continued support from federal, state, national and local governments will also continue to boost the collection, processing and demand for recycled plastic. On the downside, there are limitations in plastics recycling in major markets (e.g., construction, automotive and packaging) due to a lack of collection capability or economical processing. Also, large-scale exports from the USA and EU to countries such as China reduce the amount of post-consumer plastic available to indigenous markets. Contamination and other health-related issues are other restrictions on plastics recycling.

As the specific examples cited in this section indicate, because of its excellent performance properties and environmental credentials, PET is benefiting more than most plastics from market trends. In addition to its use as a replacement material for traditional materials (as mentioned above), another positive factor for PET is the high growth rates being experienced for PET packaging in the developing countries of Asia Pacific, South America and Eastern Europe due to growing real incomes and increasing popularity of PET bottles. There are also good growth opportunities for PET due to the increasing demand for barrier PET bottles and jars for juice, milk, tea, beer, wine and food. However, some trends are restricting its growth, particularly in tonnage terms, and these include the move towards lighter products for PET packaging (e.g., bottles) and the increasing degree to which rPET is being used in new products [23].

The management of PET and other plastic waste is a global problem and is being addressed worldwide. Darbari and Sugumar [24] reviewed the environmental practices and plastic-recycling technologies that Japan uses to convert plastic waste into value-added materials and products. They highlighted two of the important technological options that are practiced successfully: (i) recycling of post-consumer PET bottles back into new bottles *via* a closed loop system; (ii) conversion of all types of plastic waste into ammonia gas (which is then used in fertiliser production). These recycling activities have the additional benefit of reducing the burden on the environment from petroleum and mining industries. With regard to environmental impact, to assess if a new recycling process or recycling option compares well with those being used currently, a full life-cycle assessment (LCA) can be carried out using the framework and principles of accepted standards (e.g., International Organization for Standardization standard, ISO 14044).

An LCA was undertaken by Rigamonti and co-workers [25] using the EASE-WASTE model on the plastic fraction present in municipal solid waste streams. To carry out the work they defined and modeled five plastic-waste recovery routes. The five scenarios, designated P0 to P4, were:
- P0: Baseline scenario whereby the plastic is treated as residual waste and directed partly to incineration with energy recovery and partly to mechanical biological treatment.
- P1: Source separation of clean plastic fractions for material recycling.
- P2: Source separation of a mixed plastic fraction for mechanical upgrading and separation into specific plastic types, with a residual plastic fraction being down-cycled and used for 'wood items'.
- P3: A mixed-plastic fraction is source separated together with metals in a 'dry bin'.
- P4: Plastic is separated out mechanically from residual waste before incineration.

The results of that study confirmed the difficulty in clear identification of an optimal strategy for management of plastic waste. Rigamonti and co-workers found that none of the five scenarios was unequivocally the best option for all the impact categories. When moving from P0 to the other scenarios, they found that substantial improvements could be obtained for 'global warming'. For the other impact categories, the results obtained were affected by the assumptions made about the substituted marginal energy. Nevertheless, irrespective of the assumptions on marginal energy, the P4 scenario (which implied the highest quantities of specific polymer types sent to recycling) resulted in being the best option in most impact categories.

The pressure that has been applied to the packaging sector in recent years to reduce the environmental impact of its products has resulted in several LCA studies and investigations into the carbon footprint of specific products, including those that contain rPET. An example, of the latter was published by Dormer and co-workers [26] in which they described a 'cradle-to-grave study' of food trays (e.g., for mushrooms) produced from rPET. Dormer and co-workers used data provided by a plastics manufacturer to calculate the carbon footprint of these products and assess how various parameters affected this footprint. To achieve this aim, a model was developed using data on production batches. The model revealed that the cradle-to-grave carbon footprint of 1 kg of food trays containing 85% rPET was 1.58 kg of carbon dioxide. The raw material, manufacturing, secondary packaging, transport and end-of-life stages contributed 45, 38, 5, 3 and 9% of the total life-cycle greenhouse gases, respectively. The proportion of rPET in the trays was found to have a significant effect on their carbon footprint. For example, increasing the proportion of rPET to 100% resulted in a decrease of 24%. Reduction in the weight of the tray also has a significant influence, with a reduction in weight of 30% being reflected in a reduction in the carbon footprint of 28%. Changes in parameters such as transport and whether trays were recycled at end-of-life were found not to have such a significant impact, although it was evident that high recycling rates should be a goal to ensure rPET availability in the market place.

SRI Consulting have analysed the carbon footprint of PET bottles from production of raw materials to disposal, and secondary packaging from cradle to grave. In 2010, the company published the results of research which stated that unless the yield from recycling PET bottles is ≥50%, disposing of the bottles in landfill will lead to a lower carbon footprint [27]. The report also claimed that for countries with adequate space and with little recycling infrastructure, disposing of PET bottles in landfill generates a lower carbon footprint than recycling or incineration. The findings of the research also showed that take-back programmes and bottles collected through deposit programmes generate yields that are sufficiently high to produce lower carbon footprints than landfilling, but that domestic kerbside collections (Figure 1.2) do not because high contamination levels means that as much as one-third of post-consumer PET must go to low-grade applications or is lost during cleaning.

Figure 1.2: Typical example of the kerbside collection of household waste in the UK. Reproduced with permission from the Waste and Resources Action Programme, Banbury, UK. ©WRAP.

Deshmukh and Mhadeshwar [28] pointed out that the goal of sustainable polymers is to reduce the negative impact on the environment while still having a plastic material that fulfills the requirements of society. Their article highlighted PET recycling and the use of rPET in the marketplace, and did so within the context of plastics recycling in general. They provided an overview of the use of identification codes, collection and sorting of plastic waste, mechanical and chemical recycling, and use of waste plastics for energy generation. Another general review of plastics recycling that included a particular reference to PET recycling was that published by Nishida [29]. In addition to the various technologies available to rPET, that review also covered the feedstock recycling of commodity plastics, depolymerisation in supercritical and subcritical fluids, recycling of polycarbonate, reversible cross-linking/de-crosslinking polymer systems, and biomass-based recyclable polymers.

Worldwide interest in PET recycling and the important role of industry is apparent in the number of presentations being given on this subject at major conferences. For example, a paper presented by Cornell [30] to an ACS PMSE conference in 2011 reviewed recycling of post-consumer PET and discussed it in terms of the availability of raw materials, quality standards and applications for the recycled polymer. The paper had a strong focus on recycling of PET bottles and considered mechanical and chemical recycling, with the most commonly found method being glycolysis. Also covered were the difficulties presented by degradable additives in waste streams and the guidance documents that were available. At another ACS PMSE Conference, Kazuki Fukushima and co-workers [31] presented a paper that described use of a green organic catalyst for PET depolymerisation in solution and the melt; the process took days in the former, but hours in the latter. The advantages of the process included generation of clean monomers at high yields. When the new methods were compared with existing depolymerisation methods they were shown to be relatively efficient with low consumption of energy and easy accessibility. The use of organic catalysts for the chemical recycling of PET was also the subject of a paper presented by Allen and co-workers at the GPEC 2010 [32]. They described the development of a family of organic catalysts and the new recycling process for PET that they were used in. At the ANTEC 2009 conference, several papers were concerned with PET recycling. For example, a paper was presented on the improvements that could be made to the intrinsic viscosity of rPET by use of radio-frequency heating was presented by Ogasahara and co-workers [33]. Also, a paper by Michaeli and Seidel [34] on use of a melt degassing extruder to process undried rPET as an alternative to the conventional pre-drying process, with its energy, time and cost considerations, was presented. In addition, a paper by Pierre and Torkelson [35] on use of solid-state shear pulverisation (SSSP) to convert linear PET to lightly branched PET with an improvement in physical and mechanical properties, as well as a dramatic increase in the crystallisation rate which improves processability, was presented. The authors suggested that these benefits would contribute to the sustainability of PET by enabling rPET to be used more frequently for high-value applications.

As with all recycling initiatives, the challenge running alongside R&D activity is to ensure that any process that achieves its technical targets will also be economically viable and so have good commercial potential. Increases in commodity prices tend to assist in achieving this commercial goal, though the costs associated with other aspects of a recycling process (e.g., energy) can increase as well, reducing the overall net gain. A complete cost-assessment exercise, taking into account every aspect and feature of a process (e.g., capital expenditure, material costs, labour costs, energy costs, storage, packaging and transportation costs) must be carried out to provide a complete and accurate evaluation of economic viability. Of course, where a product is being sold into the marketplace, the final selling price (and a robust system to establish the final selling price and vary it over time as conditions change), are also vital for ensuring that a business is profitable and viable in the long term.

To summarise, as with all recycling activities, the challenge with recycling PET (in whatever form or product) has always been to develop processes that, ideally, have the following attributes:

- Efficient in operation with relatively low running costs
- Relatively low maintenance costs
- Capable of generating a high-quality product in a consistent manner
- Flexible design in terms of output rates
- Justifiable capital outlay
- Environmentally friendly
- Meets all requirements for health and safety
- Economically competitive with existing processes and systems
- Robust with respect to satisfying changing market requirements
- Profitable!

The degree to which a particular process, or end-use, addresses these criteria is often the defining factor in their commercial viability and success.

References

1. D.J. Hurd in *Best Practices and Industry Standards in PET Plastic Recycling*, Bronx 2000 Associates Inc., Bronx, NY, USA, 1997.
2. European Commission, COM(2014) 398 Final, Brussels, Belgium, 2nd July 2014.
3. Anon, *European Plastics News*, 2013, **40**, 10, 9.
4. *Improving Food Grade rPET Quality for Use in UK Packaging*, Final Report, Waste and Resources Action Programme (WRAP), Banbury, UK, July 2013.
5. Anon, *PETplanet Insider*, 2012, **13**, 10, 18.
6. Mergers Alliance, Plastics Europe MRG, Rexam, 2012. http://www.mergers-alliance.com.
7. Anon, *PETplanet Insider*, 2011, **12**, 3, 31.
8. D. Baker, *Retail Packaging*, 2013, Jan–Feb, 28.

9.   http://www.holfeldplastics.com/english/newsarchive/holfeld-put-the-brains-and-the-beauty-into-their-latest-unveiling-the-plus-range-of-rpet-pe-poultry-and-fish-packaging/
10. M.J. Deschamps, *European Plastics News*, 2012, **39**, 3, 22.
11. Anon, *International Bottler & Packer*, 2012, **86**, 2, 30.
12. Anon, *PETplanet Insider*, 2012, **13**, 11, 22.
13. Anon, *Advanced Packaging Technology*, 2012, **1**, 3, 3.
14. Anon, *Retail Packaging*, 2012, March–April, 8.
15. *Realising the Value of Recovered Plastics – An Update*, Market Situation Report – Summer 2010, Waste and Resources Action Programme (WRAP), Banbury, UK.
16. *UK Household Plastics Packaging Collection Survey*, RECOUP, Peterborough, UK, 2015.
17. E. Redahan, *Materials World*, 2011, **19**, 7, 9.
18. W. Date in Article 'News' Portal of Letsrecycle.com, 9th December 2013. http://www.letsrecycle.com/news/latest-news/biffa-completes-redcar-mixed-plastics-plant-upgrade/
19. S. Foster, *Materials Recycling Week*, 2008, **192**, 9, 17.
20. A. Clarke, *Plastics and Rubber Weekly*, 2011, 30th September, 1
21. http://www.fda.gov
22. American Chemistry Council, *Plastics Engineering*, 2013, **69**, 5, 34.
23. Anon, *PETplanet Insider*, 2011, **12**, 9, 14.
24. N. Darbari and S. Sugumar, *Cipet Times*, 2009, **4**, May–December, 13
25. L. Rigamonti, M. Grosso, J. Moller, V. Martiniez Sanchez, S. Magnani and T.H. Christensen, *Resource, Conservation & Recycling*, 2014, **85**, 1, 42.
26. A Dormer, D.P. Finn, P. Ward and J. Cullen, *Journal of Cleaner Production*, 2013, **51**, 1, 133.
27. M. Verespej, *Plastics News (USA)*, 2010, **22**, 25, 14.
28. S. Deshmukh and N. Mhadeshwar, *Popular Plastics and Packaging*, 2012, **57**, 6, 18.
29. H. Nishida, *Polymer Journal (Japan)*, 2011, **43**, 5, 435.
30. D.D. Cornell in *PMSE Preprints*, Ed., ACS Division of Polymeric Materials Science and Engineering, Washington, DC, USA, 2011, **105**, Fall, 1084.
31. K. Fukushima, O. Coulembier, J. Lecuyer, M.A. McNeil, P. Dubois, R.M. Waymouth, H.W. Horn, J.E. Rice and J.L Hedrick in *PMSE Preprints*, Ed., ACS Division of Polymeric Materials and Engineering, Washington, DC, USA, 2010, **102**, Spring, 111.
32. R.D. Allen, J.L. Hedrick. Kazuki Fukushima, H. Horn and J. Rice in *Proceedings of GPEC 2010 Conference*, Orlando, FL, USA, Ed., Society of Plastics Engineers, Plastics Environmental Division, Lindale, GA, USA, 8–10th March 2010, Recycling Session, Paper No.17.
33. M. Ogasahara, M. Shidou, S. Nagata, H. Hamada and Y. Leong in *Proceedings of the 67th ANTEC Conference*, Chicago, IL, USA, Ed., Society of Plastics Engineers, Brookfield, CT, USA, 22–24th June 2009, p.3056
34. W. Michaeli and H. Seidel in *Proceedings of the 67th ANTEC Conference*, Chicago, IL, USA, Ed., Society of Plastics Engineers, Brookfield, CT, USA, 22–24th June 2009, p.1188.
35. C. Pierre and J. Torkelson in *Proceedings of the 67th ANTEC Conference*, Chicago, IL, USA, Ed., Society of Plastics Engineers, Brookfield, CT, USA, 22–24th June 2009, p.629.

2 Overview of the world market for recycled polyethylene terephthalate

2.1 Introduction

The Smithers PIRA report published in 2013 [1] stated that the demand for post-consumer recycled plastic was increasing and would be driven by several factors:
1. Growing emphasis on sustainability among manufacturers of packaging and consumer products;
2. Advances in processing and sorting technologies that allow a wider variety of plastic to be recycled into high-quality resins;
3. Improved collection infrastructure that increases rates of plastic recycling; and
4. Continued support for recycling efforts from federal, state and local governments providing a significant boost to collection, processing and demand of recycled plastics.

Legislation and resource-protection programmes are major drivers for development of the recycled polyethylene terephthalate (rPET) industry. Also, as is the case with recycling of other materials, recycling polyethylene terephthalate (PET) conserves fossil fuel, reduces energy use and saves landfill space, resulting in reduced emissions of greenhouse gases. Owners of global brands are interested in using rPET to make a difference to their brands in environmental and economic aspects, and this approach also helps to generate demand.

The price of virgin plastics will always be inextricably linked to the cost of the feedstocks, the manufacturing process and market factors, such as supply and demand. The price of rPET and other recycled plastics is dependent on many factors, one of which is supply and demand. However, the other factors involve the costs associated with the collection, sorting, preparation (e.g., flaking), removal of various kinds of contamination (e.g., metal, fabric, other plastics, organics), and its conversion into a usable final product (e.g., pellets). Up until recently, because of these factors, the cost of the highest quality rPET (i.e., food-grade material) was similar to that of virgin polyethylene terephthalate (vPET). However, as the number of facilities capable of producing food-grade rPET increase and, as Chapter 6 demonstrates by the number of submissions that have been made to the European Food Safety Authority (EFSA) by companies interested in getting into this market, this should happen within a relatively short period of time, the price of the highest quality rPET is expected to decrease. The continuing improvement in the recovery rate and recycling infrastructure in many countries should assist this process by supporting the supply side. One possible problem for the rPET industry going

https://doi.org/10.1515/9783110640304-002

forward is the recent decrease in the oil price due to market forces but, as this section shows, it is not always price alone which is making recycled plastics an attractive option for the food-packaging industry and other sectors.

Section 2.2 provides an overview to the general plastics market, the market for plastics recycling and, to a limited degree, the recycling market for other materials (e.g., glass) in different parts of the world to assist in understanding of the market for rPET and how this segment fits into the context of the other segments. Section 2.3 goes on to discuss the specific market for rPET within a European/global context and, as one of the major sectors that vPET (and, consequently rPET) is used in is the packaging sector (particularly food packaging), this market features prominently within Section 2.3.

This topic is sufficiently large to have been addressed by several market report publications over the years, such as those marketed by Smithers PIRA, and some of these have been used in preparation of this section. Other important statistical information that has been useful includes that made available by bodies such as Waste Resources Action Programme (WRAP) and Recycling of Used Plastics Limited (RECOUP). The specific references that have been of use from these sources are given in the applicable sections.

2.2 Overview of the general plastics market and recycling markets

The market for plastics recycling is complex, and the price of a particular recycled plastic will be set and will vary according to the specific market that exists in any particular country for that material. As mentioned in Section 2.1, the market for plastics recycling will also vary due to global fluctuations in commodity prices and the overseas market for recycled plastics.

UK organisations such as WRAP and RECOUP UK have published a relatively large amount of up-to-date market information for this country, some of which has been used in this section. It is not possible to provide this depth of information for every country, but UK information can also be used as an indicator of how the market is developing for other developed nations and regions. The tonnage and price information that has been obtained from these WRAP and RECOUP publications is shown below in Tables 2.1–2.4.

2.2.1 Situation in the UK

In the UK, households produce ≈1.7 million tonnes of plastic packaging waste (Figure 2.1) each year and this overall figure can be broken down as shown in Table 2.1. The data in Tables 2.1–2.4 originated from the WRAP 2012 report

Table 2.1: Breakdown of plastic packaging waste in the UK.

Packaging product	Amount (tonnes)
Bottles	550,000
Non-bottle rigid plastic	450,000
Films and bags	720,000
Total	1,720,000

Reproduced with permission from *Collection and Sorting of Household Rigid Plastics Packaging*, Final Report, Waste and ResourcesAction Programme (WRAP), Banbury, UK, May 2012. ©2012, WRAP [2]

Table 2.2: Waste fractions separated by a typical MRF in the UK and their market value.

Material	Supplied to	Price (£/tonne)*	Price (€/ tonne)**
Mixed paper	Paper merchant or mill	100–109	125–136
Glass	Glass plant for remelt applications or recovered for use as an aggregate	24–31 (clear glass)	30–39
Metals	Metal recyclers	130–162	163–203
Mixed plastics	PRF***	***	***
Residual material	Landfill/mechanical biological treatment/ incineration	Cost of disposal: ~100	Cost of disposal: ~125

* Prices are averaged from figures published by letsrecycle.com in 2011
** Price calculated in September 2014 using an exchange rate of £1 = €1.25
*** Assuming that the MRF does not have its own capability to sort mixed plastics into separate waste streams
PRF: Plastic recycling facility
Reproduced with permission from *Collection and Sorting of Household Rigid Plastics Packaging*, Final Report, Waste and Resources Action Programme (WRAP), Banbury, UK, May 2012. ©2012, WRAP [2]

'*Collection and Sorting of Household Rigid Plastics Packaging*' [2]. The information that is presented in this WRAP report was derived from several sources, including the WRAP report '*Market Situation*' for Spring 2010 [3], the WRAP report '*Stockport Household Plastics Collection and Sorting Trial*' [4] and the RECOUP UK report '*Household Plastic Packaging Collection Survey*' [5].

A significant proportion of this waste packaging will be black or highly coloured, and there is a serious issue regarding sorting and separation for packaging that is

Table 2.3: Market value in the UK for different plastic waste streams in a baled form.

Material	Supplied to	Price (£/tonne)	Price (€/tonne)*
PET bales	PET recycler	328–361	410–451
HDPE bales	HDPE recycler	330–358	413–448
PP bales	PP recycler	100–200	125–250
Mixed plastic bottle bales	PRF	177–247	221–309
Mixed rigid plastic bales	PRF	100–170	125–213

* Price calculated in September 2014 using an exchange rate of £1 = €1.25
Reproduced with permission from *Collection and Sorting of Household Rigid Plastics Packaging*, Final Report, Waste and Resources Action Programme (WRAP), Banbury, UK, May 2012. ©2012, WRAP [2]

pigmented using the standard carbon black pigment as described in Chapter 5. Black plastic packaging is much more common in the rigid non-bottle stream than in the bottle stream; it is widely used, for example, in food trays. Of the estimated 1 million tonnes of rigid mixed plastics packaging in the waste stream indicated in Table 2.1 (i.e., combination of bottles and non-bottles), black plastic packaging could represent between 3–6% of this amount. For example, industry estimates are that there could be 25,000–60,000 tonnes of black packaging per annum in the UK household waste stream [6]. In addition to the plastic packaging sector, the same sorting problem is present in several other sectors (e.g., the automotive sector).

One of the initial stages in the separation of mixed waste that has been collected from households (and other sites) occurs in a materials recycling facility (MRF) (Figure 2.1), which separates the mixed waste into the specific fractions shown in Table 2.2 [2].

Usually, plastics recycling facilities (PRF) are interested in PET, high-density polyethylene (HDPE), polypropylene (PP) and polystyrene (PS). In terms of value, PET and HDPE are usually the most common and profitable plastics to recycle because they are relatively easy to sell and generate good returns. PRF often exclude polyvinyl chloride (PVC), multi-layer plastics and engineering plastics such as acrylonitrile–butadiene–styrene (which is used in many electrical items and children's toys). When the baled, mixed plastic is sent by a MRF to a PRF (Figure 2.2), the value of it is dependent on the proportion of the higher-value polymers (particularly PET and HDPE), whereas the amount of film and paper contamination that it contains can lower its value by 5–35%.

Some indicative UK prices according to WRAP for the plastics once they are separated into particular plastic types and baled up are shown in Table 2.3 [2].

The values of plastics products increase more rapidly at the reprocessing stage. Once the mixed plastic waste has been sorted into generic streams, they can be flaked and washed. Some processors also have extrusion facilities to change the

Table 2.4: UK market prices (2011) for recovered and recycled plastics in different forms.

Material		Supplied to	Price (£/ tonne)*	Price (€/tonne)**
PET	Baled PET bottles	PET recycler/ reprocessor	328–361	410–451
	Food-grade PET pellet	Plastic moulder	900–1,100	1,125–1,375
	Food-grade PET flake	Plastic moulder	750–950	938–1,188
	Baled jazz (mixed colour) PET bottles	PET recycler/ reprocessor	111–130	139–150
	Non-food-grade coloured PET flake	PET recycler/ reprocessor	600–800	750–1,000
HDPE	Baled natural HDPE bottles	HDPE recycler/ reprocessor	330–358	413–448
	Jazz (mixed colour) HDPE flake	Plastic moulder	300–500	375–625
	Food-grade natural HDPE pellet	Plastic moulder	900–1,000	1,125–1,250
	Non-food natural HDPE pellet	Plastic moulder	750–850	338–1,063
PP	Baled PP containers	PP recycler/ reprocessor	100–200	125–250
PS	Baled PS containers	PS recycler/ reprocessor	0–50	0–63
Films	Plastic films (LDPE and others)	Film recycler/ reprocessor	219– 255***	274–319
Others	Residual metals	Metal recycler	155–185	194–231

* Prices are averaged from figures published by letsrecycle.com and opinions from industry representatives
** Price calculated using an exchange rate of £1 = €1.25
*** Feedback from reprocessors indicates prices closer to zero in 2012 LDPE: Low-density polyethylene
Reproduced with permission from *Collection and Sorting of Household Rigid Plastics Packaging*, Final Report, Waste and Resources Action Programme (WRAP), Banbury, UK, May 2012. ©2012, WRAP [2]

flaked material into pellets. Plastics reprocessors who choose to incorporate this capability into their process will increase the value of the product significantly (Table 2.4), sometimes enabling them to sell the product at prices close to those for virgin polymers. The washing and extrusion process is technically challenging and expensive to operate, and a processor who can manage to achieve 70–80% of virgin price is considered to be doing well. Product yield is critical for plastic reprocessors, and losses have been reported as being ≤50% from labels and product contents as well as contamination which, at a disposal cost of ≈£100 per tonne (UK landfill charge), has a significant influence of profit margins. In the UK, reprocessors would like to see a consistently applied grading system for baled material in the UK that addresses specific end markets to help avoid large variations in the quality of supply [2].

Figure 2.1: Waste being processed at a materials recycling facility (MRF). Reproduced with permission from the Waste and Resources Action Programme (WRAP), Banbury, UK. ©WRAP.

Table 2.4 shows the polymer streams that PRF and/or reprocessors can recover from a mixed rigid household plastic packaging waste stream, and shows typical pricing (from 2011) at different stages of the material supply chain as value is added by recyclers [2].

Figure 2.2: Photograph of a bale of recovered plastic. Reproduced with permission from the Waste and Resources Action Programme (WRAP), Banbury, UK. ©WRAP.

2.2.2 Situation in Europe

There is a strong growth trend in recycling of household plastics and energy recovery in Europe, as can be seen in 2011 research data comparing rates across Europe. This data, which can be sourced in *Plastics: The Facts 2012* [7], split total recovery rates into the recycling rate and energy recovery rate. The recycling performance across European Union (EU) countries is 15–30% in most cases, with energy recovery levels showing a much wider range of 0–75%. Large rates of energy recovery enabled nine countries (Switzerland, Germany, Austria, Sweden, Denmark, Belgium, Luxemburg, Netherlands and Norway) to achieve a total recovery rate of >80% in 2011. Improving the ability to separate 'difficult' waste, such as laminates and black items, in the plastic waste stream would provide a potential way in which the proportion of plastic that is recycled into new products is increased, and the amount that is used for energy recovery is decreased, in these countries.

After the best-performing nine countries, there is quite a drop in performance, with the next best-performing country being France with a total recovery rate of ≈57%. Five countries (Lithuania, Greece, Bulgaria, Cyprus and Malta) had a total recovery rate of ≤20% in 2011.

The price of virgin commodities has an important influence on the market for the recycled versions, and the market for recycled plastics is no different.

With regard to the price of virgin plastics in Europe, an indication at any point in time is provided by sources such as the *Plastics News Europe* website [8]. For example, the approximate prices that could be obtained in this way for several plastics in February 2014 are shown in Table 2.5.

Table 2.5: Approximate market prices for various plastics in Europe.

Generic plastic type	Grade	Approximate price* (€/tonne)
PS	General purpose	1,900
PP	Homoplymer for injection moulding	1,450
LDPE	Film	1,550
HDPE	Injection moulding	1,450
Linear LDPE	Film	1,500
PVC	High quality	1,250
PET	Bottle	1,250

* Prices from February 2014
Adapted from http://www.plasticsnewseurope.com [8]

With regard to the total amount of plastic produced in Europe, Mergers Alliance [9] estimated in 2012 that ≈57 million tonnes were produced (Table 2.6).

Table 2.6: Plastic production within Europe.

Country	Amount of plastic produced (million tonnes)
UK	3.76
Italy	4.90
France	7.53
Benelux	11.3
Germany	22.67
Other European	6.44
Total	57

Reproduced with permission from Mergers Alliance, Plastics Europe Market Research Group (MRG), Rexam, 2012. ©2012, Mergers Alliance [9]
http://www.mergers-alliance.com

Of these 57 million tonnes of plastic, Mergers Alliance [9] have estimated that 46 million tonnes was converted into the sectors shown in Table 2.7.

Mergers Alliance [9] have also provided an estimate of how the ≈18 million tonnes of plastic shown in Table 2.7 are converted into packaging products and split between the various sectors shown in Table 2.8.

Table 2.7: Distribution of the 46 million tonnes of plastic converted in Europe by sector.

Sector	Amount (million tonnes)
Electrical and electronic	2.8
Automotive	3.7
Construction	9.7
Packaging	17.9*
Other	11.9

* How this is distributed among the various end markets in Europe is shown in Table 2.8
Reproduced with permission from Mergers Alliance, Plastics Europe Market Research Group (MRG), Rexam, 2012. ©2012, Mergers Alliance [9]
http://www.mergers-alliance.com

Table 2.8: End markets of packaging products.

End Market	Proportion used (%)
Food	51
Beverage	18
Cosmetics	5
Healthcare	6
Other	20

Reproduced with permission from Mergers Alliance, Plastics Europe Market Research Group (MRG), Rexam, 2012. ©2012, Mergers Alliance [9]
http://www.mergers-alliance.com

It has also been estimated by Mergers Alliance [9] that, in 2012, at the end of its life, 66% of this ≈18 million tonnes was recycled (new products or energy recovery) and 34% was sent to landfill.

Looking at how the various types of plastics shown in Table 2.5 are used for packaging products, the split in the types used in all types of packaging (i.e., food and non-food) is shown in Table 1.1 (Chapter 1). Table 1.1 shows PET having a share of 8.6% and a significant proportion of this share will be for food packaging. The future for marketing and selling of rPET for food packaging is believed to be secure [10]. During recent years, food-grade rPET has been almost equal in price to virgin food-grade PET (Section 2.3) but, as the number of food-grade rPET facilities increases, this should reduce its cost and the attractiveness of the recycled alternative will be enhanced.

2.2.3 Global situation

The recycling of plastic is a global phenomenon with participants (both generators of recyclate and users) in all geographical regions. At present, there is an overall movement of recycled plastic from western countries such as the EU and USA (where it is collected and separated) to the east, particularly South East Asia and China (where it is used in manufacturing). This situation exists due to the movement and concentration of manufacturing enterprises in this geographical region over the past 20 years due to the favourably low labour rates. However, this scenario could change due to increases in their labour rates, the rising cost of transportation, and potential legislation in areas such as the EU that might restrict the export of recovered plastic to retain resources and stimulate indigenous recycling industries.

Due to the infrastructure present in most countries, most of the plastic that is recycled originates from the packaging sector. The global packaging market by region is shown in Table 2.9 [11].

Table 2.9: Worldwide consumption of packaging by region – 2003 and 2009.

Region	Size of market (€ million)*	
	2003	2009
Western Europe	84.7	106
Eastern Europe	18	30
Middle East	9	18
Africa	8	12
North America	102	114
South and Central America	13	26
Asia	86	126
Oceania	4	6

* According to the World Packaging Organisation (data given originally in USD but converted to Euros using an exchange rate of $1 = €0.77) Reproduced with permission from www.worldpackaging.org [11]

With respect to the global recycling situation, the recycling of plastics lags behind materials such as metal and glass, but it has improved its status in recent years due to regulatory pressures, environmental influences, and improved collection regimens. For the global recycling markets that exist at present, plastic (predominantly PET) bottles are the leading source of plastic for recycling, accounting for more than half of all plastic recycled in 2012. Hence, the biggest growth opportunities for recycled plastics in packaging will come from PET, as shown in Section 2.3. The quantity of PET bottles that have been collected for recycling has increased in Europe by around 6-fold since 2000. For example, figures reported by Welle [12] show that the quantity of PET bottles collected has increased from ≈250,000 tonnes in 2000 to ≈1,600,000 tonnes in 2011.

The rate of plastics recycling is increasing around the world, with Europe at the forefront. The average rate of recycling for plastics packaging in the EU is 26%, with some countries such as Sweden and Belgium reaching 40% in 2013. In the US plastics recycling amounts to only 12.1% (Table 2.10) because the recycling industry faces several challenges. As with most other countries, recycling is also minimal in several other major plastics markets, including construction products, motor vehicles and packaging film, due to a lack of collection capability or economical processing. Export sales to China also reduce the portion of available plastic scrap material for reprocessing into recycled plastics packaging. The recycling of post-consumer and waste plastic is also limited by contamination. For these reasons, it is estimated that only about half the plastic collected for recycling in the US makes its way into manufactured products.

Table 2.10: Recovery rates for selected packaging in the US.

Packaging	Recovery rate (%)
Steel	69.0
Paper and paperboard	52.7
Aluminium	35.8
Glass	33.4
Plastics	12.1

Reproduced with permission from www.epa.gov [13]

The overall, combined recovery rate in the US for glass, metal, paper and board and plastic in 2011 was 48.3% and the overall recycling rate for these materials was 24.0%. The packaging recovery rates for selected materials in the US in 2010 are shown in Table 2.10 [13].

The value of the global recycled plastic packaging market (2010– 2018) is shown in Table 2.11 [1].

Table 2.11: Global market for recycled plastic packaging– 2011–2018 (€ billion*).

	2011	2012	2013	2014	2015	2016	2017	2018	CAGR % 2008–2018
Value	8.5	9.4	10.3	11.3		13.7	15.1	16.6	9.9

* Values for this table were given originally in USD but have been converted to Euros at an exchange rate of $1 = €0.77
CAGR: Compound annual growth rate
Reproduced with permission from M. Rao in *The Future of Sustainable Packaging*, Market Report, Smithers Pira, Leatherhead, UK, 2013.
©2013, Smithers Pira [1]

Recycled plastics are a global commodity and, because many manufactured products sold in Europe are manufactured in countries such as China, such places represent an important market for recycled plastics. EU countries like the UK export a large proportion of their recovered plastics from municipal and industrial sources to China (≈85%) [14], mainly *via* Hong Kong. The Chinese market is a key outlet for recycled plastic material because it is the world's largest user of virgin and recycled plastic, creating demand and competition for materials. China was estimated to have manufactured ≈58.3 million tonnes of plastic products in 2010.

Plastics Recyclers Europe (PRE) has estimated that EU recyclers produce just 4% of the plastic consumed across Europe. Some organisations see a ban on the landfilling and exporting of the 25 million tonnes waste plastic that Europe generates each year (predicted to rise to 30 million tonnes by 2020) as a way in which this situation can be addressed [15]. PRE has estimated that:

1. More than 10 million tonnes is placed into landfill
2. Approximately 8 million tonnes is incinerated
3. Just 6 million tonnes is recycled

With regard to the export of plastic waste, the PRE say that this has grown at an impressive rate. For example, from 1999 to 2013 it grew 5-fold and, since 2009, ≈800 containers of plastic waste a day have left European ports. The current European target for the recycling of 50% plastic packaging waste is regarded by the PRE as being too low and, in addition, they would like to see recycling targets for all types of plastic waste. Given that the number of 'aspirational' people in the world will grow from 1.9 billion to 3.5 billion by 2030, this is seen as essential as the demand on world resources increases.

China is also aiming to increase the amount of recycling that it conducts internally. There were reports that it planned to introduce a system to recycle 70% of its own major waste products (including waste plastic) by 2015 [16]. This move could signal a reduction in demand from areas like the EU, leading to a reduced export market and a corresponding reduction in prices.

2.3 Overview of the market for recycled polyethylene terephthalate

This section provides an overview of the market for rPET, and includes information that covers all forms of rPET used in food and non-food sectors. It also provides information concerning the overall market for PET, and a report by Smithers PIRA [17] has predicted that this market will see a strong compound growth rate of 5.3% to 2017. This situation will see the market value increase from 44 billion USD in 2012 to 57 billion USD in 2017. It also includes information on the market

for recovered PET and vPET. It defines recovered PET as a post-consumer PET that can exist in various forms, for example:
a) In a 'crushed' product form (e.g., bottles or trays) that will probably be baled up.
b) Converted into a flake form that has not been washed.
c) Converted into a flake form that has been washed to remove contamination.

For many years, the only way of obtaining rPET for the manufacture of food contact products was to obtain the material from a producer using an approved recycling process [e.g., approved by the EFSA or the US Food and Drug Administration (FDA)] for the treatment of post-consumer waste. Then, this rPET would often be blended with vPET to produce the food contact articles. Within the last few years, however, another route has opened up, with suppliers of PET resin producing grades that are combinations of rPET and vPET. Examples of this approach are the grades produced by leading producer of PET resins La Seda de Barcelona – with trade names of Artenius Unique and Artenius Elite – which are approved for food contact [18] (Chapter 6).

2.3.1 Situation in Europe

As mentioned in Chapter 1, at present a large amount of plastic is not being recycled in Europe and other areas of the world due to factors such as poor infrastructure and available markets. As referred to in Section 2.1, one of the other challenges that remains for plastics recycling in general (and PET recycling in particular) is to achieve effective separation and sorting of black and highly pigmented articles from mixed plastic waste streams by commercial sorting systems. At present, a significant amount of plastic must be landfilled due to these problems, but advances are being made in the research being carried out, for example, on new detector systems (Chapter 5). Another factor that restricts the amount of plastic that can be recycled is use of multilayer plastic laminates and packaging that contains a barrier layer. The SuperCleanQ FP7 research project (funded by the EU) is one example of the work undertaken to attempt to move closer to a solution to these problems (Chapter 6). From the background work that was carried out by the SuperCleanQ consortium, it has been estimated that, in 2010, ≈700,000 tonnes per year of unrecyclable black/highly coloured and barrier-modified PET (e.g., PET blended with Nylon MXD6) was generated in the EU that could not be reprocessed by existing PET recycling processes into food contact materials [19]. Access to a recycling process that could process at least a proportion of this 700,000 tonnes of post-consumer PET would generate considerable additional revenue for the PET recycling sector. The exact amount would be dependent upon factors such as the proportion recycled and the product that it was recycled into, for example, whether it was recycled into high-quality washed flake or food-grade pellet.

Data on the market value of vPETand rPET products in Europe were published by Petcore (a non-profit trade body located in Brussels) in 2013. Petcore figures showed that the equivalent of ≈65 billion PET bottles (1.64 million tonnes) were recycled in Europe in 2013, which was up by 7% on data from 2012 [20]. The Chairman of Petcore, Roberto Bertaggia, said that use of PET was rising and its ability to be recycled was a factor in its success as a packaging material of choice. Other points that were made by Petcore included:

a) ≈56% of the PET containers in circulation were collected for recycling and reclamation.

b) There is a huge disparity in rates of PET collection among EU member states.

c) Improved standardisation of collection and sorting processes would help increase recycling rates even further.

As stated previously, the recycling of PET for use in food packaging is a very important sector within the rPET market. This demand has led to a significant amount of research into more effective recycling systems. Those that are available to produce high-quality and food- grade rPET are reviewed in Chapter 6, and many of these systems are referred to as 'super-clean' because of their ability to eliminate contamination in PET, and other types of plastics (e.g., HDPE), down to very low levels. 'Super-clean' can, therefore, be used to refer to a 'food-grade' system. The number of 'super-clean' systems in Europe continues to grow, as illustrated by the number of submissions made to EFSA (Chapter 6). An assessment of the European 'super-clean'

PET recycling capacity in 2010 was made by the German organisation Fraunhofer IVV [21] and their data are shown in Table 2.12.

Table 2.12: Capacity for recycling of 'super-clean' PET in Europe.

Country/location	Capacity (tonnes)
Germany	140,000
Italy	78,000
France	77,000
UK	69,000
Austria	63,000
Netherlands	54,000
Other EU	109,000
Total	590,000

Adapted from F. Welle, *Kunststoffe International*, 2011, **101**, 10, 45.
©2011, Carl Hanser Verlag GmbH & Co. KG [21]

The collection of PET bottles in Europe is forecast to reach 2.1 million tonnes by 2017, compared with 1.68 million tonnes in 2012, according to data released by Petcore Europe and PRE. More than 60 billion bottles weighing 1.68 million tonnes were recycled in 2012, representing a growth of 5.6% compared with 2011, the two associations stated in a report [22]. This report also stated that collection rates increased to 52.3% of available bottles in 2012 and, according to the PRE, this has helped ease overcapacity at recyclers, which have an average plant utilisation of 80%. Non-food products such as fibres were the single largest end-market for rPET in 2012, but strong growth in the sheet and bottle markets are putting these three markets at a similar level. With the exception of two members, all EU member states managed to achieve rates of PET recycling above the target set by the EU Packaging and Packaging Waste Directive 94/62/EC for all plastics of 22.5% [23].

With respect to the costs of vPET and rPET in Europe, information regarding the European prices was made available by ICIS on their website in March 2014, and this information has been used in this section. The prices provided by ICIS for various forms of recovered PET and rPET are shown in Table 2.13.

Table 2.13: European prices for various forms of recovered PET and rPET.

Product	Price (€/tonne)*	Mid-point price (€/tonne)*
PET bottles – colourless (post- consumer)	420–550	485
PET bottles – mixed colours (post-consumer)	265–365	315
Recycled colourless PET flake (not washed)	880–990	935
Recycled mixed-colours PET flake (not washed)	680–765	723
Recycled food-grade pellets	1,190–1,300	1,245

* ICIS (3rd March 2014)
Reproduced with permission from www.icis.com [24]

The approximate price (€1,250 per tonne) of vPET in February 2014 is shown in Table 2.5. Several factors can influence the price of vPET:
a) Production costs for the 'up steam' raw materials *para*-xylene and monoethylene glycol.
b) Availability of vPET on the European market – this increased during 2013 due to the opening of new plants in Egypt, Belgium, Turkey and the UK.
c) Customer anticipation of how prices will move in the future.

It can be seen by a comparison of the vPET price in Table 2.5 with the price for recycled food-grade pellet shown in Table 2.12 that they are similar. This situation arises because there is a high demand in the marketplace due to the desire for owners of multinational brands to use recycled food-grade PET for food products and, in some cases, non-food items, such as customer-care products. For example, in a

Smithers PIRA report [17], the western European market for rPET was stated to have grown during the period 2004 to 2009 from 650,000 tonnes to 1.4 million tonnes, and in 2010, the collection of PET bottles in Europe rose by 7.4% to 1.5 million tonnes, respectively, with a collection rate of 48.3%. The report also stated that the estimated total capacity for mechanical reclamation in Europe was only 1.7 million. This finding illustrated that, at present levels of growth, there is a need for a significant increase in capacity, which is an encouraging fact for companies looking to enter into this market.

A report from Petcore and European Plastics Recyclers [25] also demonstrated that the amount of post-consumer PET collected in Europe is increasing: European PET collection reached 1.45 million tonnes in 2010, an increase of 6.5% on 2009. The collection rate of all PET bottles on the market in Europe remained at 48.3%, and 0.25 million tonnes of rPET was used to produce containers. The report also highlighted an increase in the amount of rPET re-used within Europe, as illustrated by the fact that exports to the Far East fell for a second year to 13% of the total amount collected. Also, imports of baled PET post-consumer bottles from outside Europe were lower during this period.

2.3.2 Global situation

The global market for PET recycling is estimated at €3 billion and expected to grow at 8% per year, doubling by 2020. This scenario is being encouraged by multinational companies such as Coca-Cola and Nestle, who are encouraging the move towards use of 100% rPET in food contact products. For example, in 2013, Invicta Plastics claimed that they had become the first company in the world to create rigid, food-safe products (e.g., cups) from 100% rPET [26]. It was reported that this development was attracting the interest of companies such as Coca-Cola Enterprises, Greenpac and Asda. Also, Phoenix Technologies announced in 2013 that they had launched LNOTM resin, a new food-grade rPET pellet that could be used at levels ≤100% for the manufacture of food packaging [27].

A Smithers PIRA Report [17] has described the global market, and contains the facts and predictions shown in Tables 2.14–2.16. PET is used throughout the world for the production of a wide range of packaging materials, mainly for food products (Chapter 6). The factors that drive PET demand in each end-use sector are influenced by the alternative packaging materials and pack types that PET is competing with in each segment of the market. This information is shown in Table 2.14. For several years, it was the replacement of glass that was a driver for growth in the use of PET for food packaging. However, developed countries have now largely exhausted this potential. Hence, attention is shifting to developing countries in which PET can be used as a good barrier packaging for beer, wine, juices and dairy products [28].

Table 2.14: Packaging materials that compete with PET.

Material	Water	CSD	Juices	Beer	Dairy drinks	Ready to drink teas	Other drinks	Food bottles and jars	Non-food bottles and jars	Thermoforming
Glass	*	*	*	*	*	*	*	*	*	
Cans		*		*			*			
PVC	*							*	*	*
HDPE			*		**				*	
PP							*	*	*	**
PS										**
Cartons			**		**	**				

* used; and
** predominantly used
CSD: Carbonated soft drinks
Reproduced with permission from D. Platt in *The Future of Global PET Packaging to 2017*, Smithers PIRA, Leatherhead, UK. ©2012, Smithers PIRA [17]

Table 2.15: Tonnage for the worldwide market of PET packaging by end-use sector (,000 tonnes).

Packaging market sector	2007	2012 (projected)	CAGR (%) 2007–2012	2017 (forecast)	CAGR (%) 2012–2017
Total drinks bottles	10,237.0	12,021.2	3.3	15,443.4	5.1
Food bottles and jars	814.2	988.4	4.0	1372.9	6.8
Detergents and cleaning products	303.2	339.8	2.3	417.8	4.2
Cosmetics and toiletries	141.0	162.9	2.9	205.5	4.8
Pharmaceutical and medical	225.7	267.3	3.4	352.1	5.7
Non-food bottles and jars	669.9	769.9	2.8	975.4	4.8
Thermoforming	852.6	993.6	3.1	1,274.5	5.1
Total	12,573.7	14,773.1	3.3	19,066.2	5.2

Reproduced with permission from D. Platt in *The Future of Global PET Packaging to 2017*, Smithers PIRA, Leatherhead, UK. ©2012, Smithers PIRA [17]

Rapid growth in the use of PET in these areas can be illustrated clearly. In 2010, PET, glass, cartons and cans together accounted for 86% of global packaging for soft drinks but, by 2012, PET alone accounted for 61% of the global market for soft drinks, mainly at the expense of glass. Some of the advantages of PET in this market

Table 2.16: Tonnage for the worldwide market of PET packaging by geographical region (,000 tonnes).

Area	2007	2012 (projected)	CAGR (%) 2007–2012	2017 (forecast)	CAGR (%) 2012–2017
Western Europe	2,816.8	2,903.6	0.6	3,219.2	2.1
Central and Eastern Europe	1,048.6	1,190.7	2.6	1,552.2	5.4
Middle East and Africa	1,157.2	1,391.7	3.8	1,864.8	6.0
North America	3,411.1	3,564.3	0.9	4,138.1	3.0
South and Central America	1,131.9	1,383.9	4.1	1,826.8	5.7
Asia Pacific	3,008.1	4,338.8	7.6	6,465.0	8.3
Total	12,573.7	14,773.1	3.3	19,066.2	5.2

Reproduced with permission from D. Platt in *The Future of Global PET Packaging to 2017*, Smithers PIRA, Leatherhead, UK. ©2012, Smithers PIRA [17]

are that PET bottles are easier for retailers to handle, do not break, can be resealed, and are light for on-the-go use by consumers. Advances in technology (e.g., barrier layers, oxygen scavengers, ultraviolet blockers) that have addressed shelf-life and other issues have also assisted PET in its growth.

The information shown in Table 2.15 confirms that the market for PET packaging is very buoyant, and that all sectors are predicted to grow at an increasing rate in the immediate term. The market for food-grade rPET is, therefore, expected to be strong due to the increasing interest in using PET for all packaging applications.

In the same Smithers PIRA report, the market tonnages were broken down into different geographical regions and this information is shown in Table 2.16.

Several authors have reviewed consumption and recycling in specific countries outside Europe. For example, Schmitke [29] described how PET consumption in Brazil 2014 was 600,000 tonnes per year and was forecast to reach double-digit growth rates. With regard to the recycling rate of PET, Schmitke stated that this was estimated to be ≈54%. Also, in an article published in the same journal in 2013, Roscheck [30] reported that ≈76.5% of the volume of soft drinks in Uruguay was packaged in PET in 2012. Also mentioned in the article was that the sales of soft drinks in Uruguay were said to have grown from 400 million litres in 2007 to 490 million litres in 2012, and that the major South American producer of PET preforms, Cristalpet, was preparing for an increase in the amount of rPET that it processed.

In a regional market report on China by von Schroeter [31], interviews were featured with representatives of four companies with a major interest in PET: Krauss-Maffei Group China, S.O.E. Group, Xiamen Yinlu Foods and KHS Corpoplast. One of the topics covered during the interviews was PET recycling. The S.O.E. Group (said to be a converter which supplies vPET preforms) was reported as having been interested in using rPET for several years. Hence, it had held talks with the local municipalities about collection systems, and with the Chinese authorities over the use of rPET in food applications.

With respect to approvals for new food-grade PET recycling processes in the US (Section 4.3.4), a process for producing PET food packaging from 100% rPET bottle flakes was approved by the FDA in 2010 [32]. This process was developed by Gneuss using a combination of a multiple-rotation extruder, melt-filtration system, and a solid-state polycondensation process to increase the viscosity of the final PET product to the required level. The process is claimed to have several advantages over conventional recycling processes: relatively low investment, low energy costs and enhanced quality of the product. The FDA approval is for the pellets from the process to be used with hot- and cold-filled PET bottles.

Bottle-to-bottle recycling is well established in the US, where the FDA has established acceptable approaches for the use of rPET in bottle applications. These approaches involve use of a surrogate compound testing procedure (i.e., challenge testing) for the recycling process, which is closely related to the EFSA marker compound test used in the EU (Chapters 4 and 6). The Smithers PIRA report [17] states that the recycling rate for PET containers in the US increased 7 years in a row to stand at 29% in 2010, and that the amount of rPET that was re-used was also at its highest value (454,000 tonnes), with applications for food and drink representing 21.6% of this value. The report also states that >50% of the PET that was collected in the US was exported to China.

Some of the many commercially available recycling systems (Chapter 6) for the production of food-grade rPET have achieved greater market penetration than others. It is claimed that the Vacurema technology owned by Erema is the clear global market leader, with a 50% share in the rPET direct food contact market [33]. For example, in the US, of the 779,000 tonnes of PET collected each year, 295,000 is processed into rPET for direct food contact, and 50% of this is done using Vacurema technology.

Another aspect of the global food-grade rPET market concerns the setting up of 'in-house' systems within large food manufacturing factories. These set ups (in which bottles containing rPET are manufactured on the same site where the food product is manufactured) are economically and environmentally attractive because they remove the need to transport large numbers of empty bottles to a site for filling. An example is the in-house bottle manufacturing facility at the GSK Lucozade and Ribena factory in Coleford, Gloucestershire, in the UK. Logoplaste operate the facility, which converted >15,000 tonnes of vPET and 4,000 tonnes of rPET into bottles in 2012 [34].

At present, a positive outlook exists for further global growth in the use of rPET for new products. However, particularly in the food- packaging sector, there are areas of potential concern that have been commented upon. For example, to be used for food contact packaging, rPET must meet regulations on food contact (e.g., the Plastics Recycling Regulation, EC 282/2008 and the Plastics Regulation EU 10/2011 in the EU – Chapter 4). However, this is a minimum requirement as far as certain sectors of the food industry are concerned. Several brand owners are much stricter in the criteria that they use in their decision-making, and not all commercial PET recycling processes that meet these EU regulatory requirements will be acceptable to them [35]. This fact must be considered when assessing the commercial potential for new processes for PET recycling because this will depend upon its acceptance by industry as well as its technical capabilities. In 2012, *PETplanet Insider* reported [36] that APPE (the Packaging Division of Las-Seda de Barcelona) stated that a lack of high-quality food-grade rPET was hampering the aspirations of brand owners and product fillers to increase the amount of rPET in their packaging.

This concern with regard to quality, as well as limitations on the available supply of food-grade rPET, has also been raised in other sections of the trade press [37]. In this article, it was stated that concerns in these areas were tempering expectations of dramatic growth for rPET in food packaging, despite commitments by Coca-Cola and Evian Volvic to use the material.

As well as contamination resulting from polymers other than PET (as well as pigments and additives within these products), increasing levels of recycling in general can cause their own problems. For example, it has been reported [38] that food containers made with recycled plastic from international suppliers have been found to be contaminated with heavy metals (e.g., cadmium, antimony, nickel, chromium and lead) due to the recycling and sorting processes that have been employed. It is suspected that these heavy metals had originated from plastic flake becoming contaminated with electronic waste.

Another possible restraint on the use of more rPET in certain types of food packaging has been the subject of a paper at a recent conference [39]. The presentation commented that, although it is now common to see 25–30% rPET in PET bottles for carbonated soft drinks, there are concerns that increasing this amount could cause a reduction in their performance. This hypothesis arose because some studies showed that bottles with high rPET content, although stiffer and tougher than vPET bottles in the axial direction, are softer and weaker in the hoop direction and have greater creep. These weaknesses could affect the ability of the bottle to retain its shape and so affect its stacking stability, vending and overflow capacity. Despite these concerns, there are PET bottles in commercial use for carbonated drinks that contain >40% rPET. An example is a 1.5-l bottle for Lidl which, in addition to having relatively high rPET content, is also one of the lightest bottles (26.8 g) in the world. Also, for the North America market, PepsiCo are reported to have created the first plastic drink bottle (7UP EcoGreen) from 100% rPET [40].

References

1.   M. Rao in *The Future of Sustainable Packaging*, Smithers PIRA, Leatherhead, UK, 2013, p.54.
2.   *Collection and Sorting of Household Rigid Plastics Packaging*, Final Report, Waste and Resources Action Programme (WRAP), Banbury, UK, May 2012.
3.   *Market Situation Report: Realising the Value of Recovered Plastics*, Waste and Resources Action Programme (WRAP), Banbury, UK, Spring 2010.
4.   *Stockport Household Plastics Collection and Sorting Trial*, Final Report, Waste and Resources Action Programme (WRAP), Banbury, UK, 2005.
5.   *UK Household Plastics Packaging Collection Survey 2011*, Recycling Of Used Plastics Limited (RECOUP), Peterborough, UK, 2011.
6.   *Development of NIR Detectable Black Plastic Packaging*, Waste and Resources Action Programme (WRAP), Banbury, UK, September 2011.
7.   *Plastics – The Facts 2012: An analysis of European Plastics Production, Demand and Waste Data for 2011*, Plastics Europe, Association of Plastics Manufacturers, Brussels, Belgium, 2012.
8.   http://www.plasticsnewseurope.com
9.   Mergers Alliance, Plastics Europe Market Research Group (MRG), Rexam, 2012. http://www.mergers-alliance.com.
10. Anon, *Bioplastics World*, 2013, **1**, 10, 4.
11. http://www.worldpackaging.org
12. F. Welle, *Resources, Conservation and Recycling*, 2013, **73**, 1, 41.
13. www.epa.gov
14. *The Chinese Markets for Recovered Paper and Plastics – An Update*, Waste and Resources Action Programme (WRAP), Banbury, UK, Spring 2009.
15. http://www.plasticsrecyclers.eu
16. http://www.packagingnews.co.uk
17. D. Platt in *The Future of Global PET Packaging to 2017*, Smithers PIRA, Leatherhead, UK, 2012.
18. Anon, *PETplanet Insider*, 2012, **13**, 7–8, 31.
19. SuperCleanQ EU-funded FP7 Research Project. http://www.supercleanq.eu
20. http://www.petcore.org
21. F. Welle, *Kunststoffe International*, 2011, **101**, 10, 45.
22. *Post Consumer PET Recycling in Europe 2012 and Prospects to 2017*, PCI PET Packaging, Resin and Recycling Ltd, Speldhurst, Kent, UK.
23. http://www.plasteurope.com, Article published on 16th August 2013
24. http://www.icis.com
25. Anon, *PETplanet Insider*, 2011, **12**, 10, 16.
26. Anon, *British Plastics and Rubber*, 2013, March, 9.
27. Anon, *PETplanet Insider*, 2013, **14**, 6, 24.
28. F. Welle, *Kunststoffe International*, 2013, **103**, 10, 64.
29. W. Schmitke, *PETplanet Insider*, 2014, **15**, 1–2, 10.
30. F. Roscheck, *PETplanet Insider*, 2013, **14**, 12, 10.
31. W. von Schroeter, *PETplanet Insider*, 2012, **13**, 10, 10.
32. Anon, *High Performance Plastics*, April 2010, p11.
33. Anon, *Popular Plastics and Packaging*, 2014, **59**, 3, 53.
34. Anon, *Plastics and Rubber Weekly*, 2013, 4th October, p.23.
35. Anon, *PETplanet Insider*, 2011, **12**, 3, 20.
36. Anon, *PETplanet Insider*, 2012, **13**, 3, 10.

37. Anon, *Advanced Packaging Technology World*, 2013, **1**, 6, 2.
38. Anon, *Journal of Plastics Film and Sheeting*, 2013, **29**, 2, 63.
39. J.Z. Yuan, C.A. Hayes and P.A. Harrell in *Proceedings of the 71st ANTEC Conference*, Cincinnati, OH, USA, Ed., Society of Plastics Engineers, Brookfield, CT, USA, 22–24th April, 2013, Paper No.1584107, p.4.
40. Anon, *PETplanet Insider*, 2012, **13**, 1–2, 26.

3 Brief history of the recycling of polyethylene terephthalate

3.1 Introduction

This section provides a brief overview of some of the important developments, events and milestones that have occurred in the recycling of polyethylene terephthalate (PET). It does not attempt to describe these items in chronological order, but instead groups them into individual sections in which the contents of each section share a degree of commonality, be it geographical or technical. The history of any recycling industry can be charted and described by numerous statistics that quantify various aspects of its' activities and the industry for PET recycling is no different. Statistical information is laid out in this section, but a large amount of statistical information shown in Chapter 2 is pertinent to the historical development of this sector and can be used to gauge its success against several criteria (e.g., economic and environmental), and the degree of influence that it now has in our daily lives. There is also historical information in other sections of this book. The information presented below can be used in conjunction with this information to provide a fuller picture of the development and progress of the PET-recycling industry over the last 40 years.

PET recycling has been at the vanguard of post-consumer plastics recycling in several countries and regions to a significant extent. One of the first plastics products to be collected and recycled in significant quantities were plastic bottles, particularly food-grade bottles, and most of these are made from PET. Today, plastic bottles are the leading source of plastic for recycling, and accounted for more than half of all plastic recycled in 2012 (Chapter 2). The quantity of PET bottles that has been collected for recycling has increased in Europe by approximately 6-fold since 2000. For example, figures reported by Welle [1] show that the quantity of PET bottles collected has increased from ≈250,000 tonnes in 2000 to ≈1,600,000 tonnes in 2011 (Section 2.2). The rate of plastics recycling is increasing worldwide (with Europeat the forefront) but it lags behind recycling for other materials, particularly metal and glass. Plastics recycling amounts to only 12.1% in the US (Table 2.10) because their recycling industry faces many challenges. Recycling is also minimal in several other major plastics markets, including construction products, motor vehicles and packaging film, due to a lack of collection capability or economical processing. Export sales to China also reduce the portion of available plastic scrap material for reprocessing into recycled plastics packaging. The recycling of plastics is also limited by factors such as contamination and other health issues. It is estimated that only about half of the plastic collected for recycling in the US makes its way into manufactured products.

PET recycling should be set into the general context of the recycling of plastics in general, and mention has already been made of major initiatives such as the

https://doi.org/10.1515/9783110640304-003

Integrated Product Policy in Europe (Section 4.1). To reach the goal of sustainability (defined as a creating a balance between technology, economics and environmental issues) a self-supporting recycling infrastructure must be created for the plastics industry as a whole. To achieve this noble aim, several features are required [2]:

– Cost-effective technologies for sorting and recycling:
  – Need investment in new recycling technologies
  – Target areas of greatest need and successful transfer to industry
– Design of products that can be recycled easily (design for recycling) [3]:
  – Reduce the number of plastic types and mark parts
  – Avoid combining incompatible plastics in a component/product
– Design of products that can be disassembled easily (design for disassembly):
  – Ensure easy and cost-effective separation of different components
– Creation of markets for recycled plastics – issues to address:
  – Lack of knowledge on consistency of quality and properties
  – Price, price fluctuations and consistency of supply
– Creation of an effective recycling infrastructure:
  – Ensure that post-consumer waste is reprocessed in a timely manner
  – Encourage all involved in waste recycling/management through education
  – Encourage purchase of goods containing recycled material
– Quality must be assured to maintain confidence in products:
  – Standardisation programmes, such as those championed by the Waste and Resources Action Programme (WRAP) and Consortium for Automotive Recycling(CARE) in the UK to help achieve this aim

As discussed in Chapter 2, due to the infrastructure present in most countries, most of the plastic that is recycled originates from the packaging sector. The average rate of recycling for plastic packaging in the European Union (EU) is 26%, but some countries, such as Sweden and Belgium, reached 40% by 2013. An image showing a collection of post-consumer plastic products in shown in Figure 3.1, and a plastic waste stream being processed is shown in Figure 3.2.

The amount of plastics recycled worldwide lags behind the overall global recycling market for traditional materials, such as metal and glass, but it has improved its status in recent years due to the pressures and influences described in Chapter 1 and technological advances, such as improved collection regimens, sorting techniques and recycling processes. Continued support from governments also continues to boost the collection, processing and demand for recycled plastic.

3.2 Situation in Europe

As the need to increase recycling rates in economic areas such as the EU has increased, different national governments have adopted different measures to try and

Figure 3.1: A collection of post-consumer plastic products. Reproduced with permission from the Waste and Resources Action Programme (WRAP), Banbury, UK. ©WRAP.

Figure 3.2: Example of a plastic waste steam being processed. Reproduced with permission from the Waste and Resources Action Programme (WRAP), Banbury, UK. ©WRAP.

achieve this result. Some of the measures that have been taken over the last 20 years, and the countries that they have been used in, are listed below [2].
- Increase in price of virgin plastics (Italy):
  - Increases use of recycled plastics
- Charges for waste disposal (UK):
  - Increasing charges for landfill encourages recycling

- Charges for waste production (Sweden):
  - Encourages use of re-usable materials
- Charges for waste collection (Belgium):
  - Encourages recycling and reduces waste generation
- Deposit-refund systems (Germany):
  - Encourages return of post-consumer products for recycling

To recycle plastic, it must first be collected. In several countries (e.g., the UK), it is only within the last decade that a significant proportion of the households in the country have been covered by a 'kerbside' collection system for plastic. Initially, in the UK, due to the ubiquitous use of PET for the manufacture of carbonated and non-carbonated beverages, low level of contamination, and ease of identification, the only plastic items collected in many regions were PET bottles. Within the last 5 years this situation has changed and several other types of plastic packaging waste are now being collected. Data on the amount and specific types of plastic being collected in the UK and other countries is provided in Chapter 2.

An article published in *Plastic Packaging Innovation News* in January 2007 [4] illustrates how relatively recently the widespread introduction of recycled polyethylene terephthalate (rPET) packaging onto retailer's shelves has occurred. This article reported that, by the end of quarter 1, 2007, the UK retailers Marks & Spencer, Tesco and Sainsbury's will all have post-consumer PET packaging on their shelves. It was reported in this article that the retailers were intending to use the packaging of RPC Bebo UK, with Marks & Spencer using rPET lids for some bottles and Sainsbury's changing to rPET styling trays. At the time, RPC were reported as saying that obtaining sufficient rPET in the UK to meet demand was a challenge because ≈70% of it was being exported to China. In 2006, it was reported in the same journal that Coca-Cola Enterprises (a UK franchise of Coca-Cola) had started to use rPET packaging containing between 25 and 50% rPET for its 500-ml bottles. At the time, the company was reported as having to choose between three post-consumer rPET resins for these bottles, and it eventually chose Supercycle from Amcor [5].

A measure of how things have changed in Europe over the last 15 years is provided by reference to historical facts and figures for recycling in this region that show an upward trend and which can be compared with similar data shown for the US in Tables 3.2 and 3.3. One of the principal sources for this type of data in the US is the National Association for PET Container Resources (NAPCOR) and the Association of Postconsumer Plastic Recyclers (APR). For Europe, PET Containers Recycling (Petcore) [6] and Plastics Recyclers Europe (PRE) [7] provide a similar service. Welle used figures made available by Petcore in a review paper entitled '*Twenty Years of PET Bottle-to-Bottle Recycling – An Overview*' [8] which presents some statistics that relate to the number of bottles collected in the EU from 1995 to 2009. Welle showed that in 1995 the US was ahead of the EU, with ≈350,000 tonnes being collected as opposed to <50,000 tonnes in the EU, but that by 2009 the collection rate in the EU stood at ≈1,350,000 tonnes, whereas in the US it was only ≈650,000 tonnes. Some of the other information in the paper provided by Petcore is shown in Table 3.1. This information relates to the end markets for PET recyclates in the EU from 2001 to 2009 and shows a distinct trend away from fibre products, towards bottles and, to a lesser extent, sheet products.

Table 3.1: End markets for PET recyclates in the EU – 2001 to 2009.

| Year | Product category (%) | | | | |
	Fibre	Sheet	Bottle (food and non-food)	Strapping	Other
2001	62.0	16	7	10	5
2005	57.0	16.2	15.2	7.8	3.8
2009	40.5	27.0	22.0	7.0	3.5

Adapted from http://petcore.org [6] and http://www.napcor.com [9], 2011

There are a lot of historical data on the recycling rates for PET bottles because of their importance to the food-packaging sector and recycling industry. Other milestone announcements by Petcore include the one in 2003 [10]: more PET bottles were being recycled in Europe than ever before, with an increase of 36% over the 2002 figure being recorded to bring the total for the year to 612,000 tonnes. Growth rates were high for several countries, including Germany and the UK, but it was noted at the time that these two countries were also contributing most to the increase in bottle exports to China. Petcore also reported that the markets for rPET remained substantially unchanged, with bottle-to-bottle (B2B) recycling increasing from 8.1% in 2002 to 11.1% in 2003, polyester fibre at 70.4%, sheet at 7.5%, and strapping at 7.6%. These 2002/2003 figures can be compared with figures published by Petcore Europe and PRE in 2012 (Section 2.3.1), which showed that 60 million bottles weighing 1.68 million tonnes were recycled, a significant increase. Looking ahead, it was reported on the Plasteurope website [11] that data released by Petcore Europe and Plastics and Recyclers Europe in a report compiled by PCI PET Packaging, Resin & Recycling Ltd [12] predicted that the collection of PET bottles in Europe would reach 2.1 million tonnes by 2017, which compares with the figure of 1.68 million tonnes in 2012 referred to above.

Further statistical information on the recycling situation in Europe was provided in an article in *PETplanet Insider* [13]: Petcore and PRE recorded that European post-sorting PET collection had reached 1.45 million tonnes in 2010, an increase of 6.5% on 2009. Also, the overall collection rate in 2010 remained at 48.3% of all PET bottles on the market, and around one-quarter of a million tonnes of this material was used to produce new bottles. In the same article it was stated that ≈13% of collected PET was exported to the Far East (which represented a decline) and that the amount of imported baled PET bottles from outside Europe also declined. Petcore/PRE are also reported as saying that the amount of exported plastic waste has grown enormously. For example, from 1999 to 2013 it grew 5-fold, and this scenario can be regarded as a problem because it has deprived European recyclers of the raw materials they need to develop and grow their businesses. This situation may start to change in the coming years as countries like China start to recycle more of their own plastics (a target of 70% by 2015 had been set by China), and so they might be less prepared to accept some of the waste plastic they have imported to date. Finally, the current European target for the recycling of 50% plastic packaging waste is reported as being regarded by the Petcore/PRE as being too low, and that they would like to see recycling targets for all types of plastic waste. Given that the number of 'aspirational people' in the world is going to grow from 1.9 to 3.5 billion by 2030, this move is seen as essential by the organisations as the demand on the world's resources increases.

Today, there is a greater emphasis than ever in the EU on plastic recycling, as demonstrated by the European Commission's recent green paper, which indicated that revised legislation affecting this sector will come into being within the next few years [14]. The main focus of this legislation is likely to be greater emphasis on recycling targets, solving the problem of the landfilling of plastics and increasing the quality of recyclates. By restricting (or even banning) the landfilling of the 25 million tonnes of waste plastic that Europe generates each year (predicted to rise to 30 million tonnes by 2020), it is hoped that the indigenous plastic-recycling industry will be supported and encouraged to grow from a relatively low base. For example, the PRE organisation has estimated that EU recyclers produce just 4% of the plastic consumed across Europe. This organisation also estimates that:

1. More than 10 million tonnes is placed into landfill
2. Approximately 8 million tonnes is incinerated
3. Just 6 million tonnes is recycled

3.3 Situation in the US and other parts of the world

In the US, ≈12% of plastic packaging was recycled according to an assessment by the US Environmental Protection Agency in 2010 [15] (Table 2.10). With regard to the recycling markets that exist at present, plastic bottles are the leading sources of plastic for recycling, accounting for more than half of all plastic recycled in 2012. The biggest growth opportunities for recycled plastics in the packaging sector will come from PET, in bottle form and other types of packaging. With regard to these opportunities for PET recycling in the US, *PETplanet Insider* [13] has stated that a NAPCOR/APR report found that the total rPET production in 2009 was 561 million lbs, with a further 601 million lbs in clear flake being exported. The report also found that the use of rPET in food, beverage and non-food PET containers increased by 37% from 2008 to 2009.

A more recent NAPCOR/APPR report published in 2015 [16] showed that the amount of PET available for recycling in the US was increasing. Specifically, it stated that the total weight of PET bottles and jars available in the US for recycling in 2014 was 5,849 million lbs, an increase of 1.5% over 2013. The total amount of post-consumer PET bottles collected and sold for recycling in 2014 was 1,812 million lbs, representing a recovery rate of 31.0%. The report also provided the gross recycling rates for PET bottles from 2004 to 2014 and these figures are shown in Table 3.2.

Table 3.2: Gross recycling rates – 2004 to 2014.

Year	Total bottles collected in the US (million lbs)	Bottles on shelves in the US (million lbs)	Gross recycling rate (%)
2004	1,003	4,637	21.6
2005	1,170	5,075	23.1
2006	1,272	5,424	23.5
2007	1,396	5,683	24.6
2008	1,451	5,366	27.0
2009	1,444	5,149	28.0
2010	1,557	5,350	29.1
2011	1,604	5,478	29.3
2012	1,718	5,586	30.8
2013	1,798	5,764	31.2
2014	1,812	5,849	31.0

Reproduced with permission from Report on Postconsumer PET Container Recycling Activity in 2014, NAPCOR/APR, Napor, 13th October 2015. ©2015, Napor [16] http://www.napcor.com/PET/pet_reports.html

With regard to the number of bottles collected in Table 3.2 purchased by US re-claimers, this amounted to 77% of all bottles, which was up from 74% in 2013. US reclaimers were also reported to supplement their domestic purchases by importing 177 million lbs of post-consumer bottles or 'dirty' flake mainly from Canada and Mexico as well as Central and South America. Once other sources of PET are in-cluded, such as post-consumer thermoforms, post-consumer strapping and unpro-cessed industrial scrap, the total quantity of scrap PET that US reclaimers purchased in 2014 was 1,660 million lbs.

In addition to data concerning specifically the recycling of PET bottles, the NAPCOR/APR 2015 report also addressed the use of rPET in a wide range of prod-ucts. It contains figures that show how much rPET was used in these products from 2004 to 2014 [16] (Table 3.3).

Taken as a whole, the data presented in Table 3.3 show that the amount of rPET used in all sectors showed a steady increase year-on-year from a total value of 878 million lbs in 2004 to 1,564 million lbs in 2014.

Table 3.3: rPET used by product category.

	Product category (amount in million lbs)						
Year	Fibre	Sheet and film	Strapping	Engineering resin	Food and beverage bottles	Non-food bottles	Other
2004	479	58	116	12	126	63	24
2005	463	71	131	8	115	63	13
2006	422	74	132	9	139	49	30
2007	383	128	144	11	136	60	38
2008	391	153	137	7	141	55	31
2009	344	159	114	10	203	65	42
2010	381	195	127	9	216	58	16
2011	398	202	120	Included within other value	242	57	21
2012	512	307	136	Included within other value	276	50	31
2013	558	315	140	Included within other value	425	50	25
2014	638	365	126	Included within other value	351	57	27

Reproduced with permission from Report on Postconsumer PET Container Recycling Activity in 2014, NAPCOR/APR, Napor, 13th October 2015. ©2015, Napor [16] http://www.napcor.com/PET/pet_reports.html

In addition to the developments in PET recycling taking place in Europe and America, it was reported in an article in 2012 [17] that the PhoenixPET food-grade rPET resin from Extrupet had been launched in Africa. The article also explained that Extrupet became the first food-grade rPET B2B recycling plant in Africa in 2009 and, at the time of the article, was the only B2B recycler on the continent. The company uses Vacurema technology from Erema (Chapter 6) to create the PhoenixPET resin product that can be used for bottle manufacture and thermoformed sheet food contact products.

Looking at other parts of the world, it has been reported [18] that Japan has one of the most efficient collection systems for PET bottles in the world, with ≈72.1% being sent for recycling. A recent development is that the Japanese FP Corporation has been pursuing bottle-to-food-grade tray recycling. In addition, in late-2010, the first B2B recycling system from Krones went into operation at a clients' facility in Gifu-Hashima, enabling food-grade recycled products to be generated.

3.4 Use of recycled polyethylene terephthalate in food contact products

The use of virgin polyethylene terephthalate (vPET) for packaging food continues to increase and, therefore, given that rPET can be produced in a high-quality, food-grade form (Chapter 6), this should have a positive effect on the amount of rPET used in this sector. Data showing how the market for PET and rPET continues to grow is shown in Chapter 2. Another illustration of how specific markets for this material are expanding and how it is being used as a replacement for traditional food-packaging materials was provided in a presentation by Maiseviciute [19]. Maiseviciute predicted that 43% of carbonated drinks would be in PET bottles by 2014 and that within a few years (at current trends) PET would overtake metal cans for packaging of carbonated drinks. rPET is very capable of being used at high replacement levels for vPET, in fact in quantities of ≤100% since 2012 (see below and Chapters 2 and 8).

The use of large amounts of rPET in food contact products has been taking place for several years. For example, in 2001 it was announced in an article in *British Plastics and Rubber* [20] that full-scale production had started in the US of a food contact bottle made from 100% post-consumer PET. The bottle, which was to be used for packaging water, was made by Plastics Technologies, the first company to obtain approval by the US Food and Drug Administration (FDA) for the production of food contact products from rPET. Recycling of rPET was carried out by a subsidiary company, Phoenix Technologies, which claimed to be the largest pelletiser of rPET in the world and had a capacity at that time of 22,000 tonnes of non-food-grade rPET and 7,500 tonnes of food-grade rPET. The FDA started issuing 'letters of no objection' to the owners of recycling processes that could produce food-grade rPET in the early-1990s and, by the time of writing, there are 187 such letters listed on the FDA website, most for the production of food-grade rPET (Section 4.3.3).

As shown above, new non-carbonated bottles containing 100% rPET have been available for ≈15 years but, due to technical considerations, it has taken longer in the case of carbonated drinks. However, progress has been made over the years, with the amounts of rPET that can be introduced into new food-grade, carbonated PET bottles increasing. This progress culminated in an announcement in 2012 [21] that PepsiCo had introduced the first soft drinks bottle, called 7UP EcoGreen, made from 100% rPET, in the US. This product has also been the subject of an article in the *European Plastics News* [22] in which it was mentioned that this bottle was intended for diet and regular types of carbonated drink sold in Canada, and was expected to reduce the amount of vPET used for those products by 6 million lbs a year. To put the developments mentioned above into context, it has been reported that PepsiCo have also developed the first PET bottle made entirely from plant-derived, fully renewable sources [23]. This bottle, which is manufactured from bio-based raw materials, includes switch grass, pine bark and corn husks, and is totally recyclable. In the same article, it was reported that in the future the company expects to broaden the range of renewable sources to include orange peels, potato peels, oat hulls and other agricultural byproducts from its food business. The bottle was expected to go into pilot production in 2012.

In addition to the use of increasing amounts of rPET in PET bottles, there has also been a trend to reducing the weight of PET bottles and so reduce the amount of

PET that ends up in the waste stream at the end of its life. For example, PET bottles used by the company Lidl in its own Freeway and Saskia brands are manufactured using >40% of rPET. Krones have proved that it is possible to reduce the weight of these bottles by ≈30% over the last few years while maintaining their functionality and handling qualities [24].

As well as bottles, the amount of rPET in rigid food contact products has been increasing. In 2013, a British company, Invicta Plastics, was reported as claiming [25] that it had beaten global competition to create the first rigid, food-safe products from 100% post-consumer PET bottles. The company also claimed that it had manufactured food-safe products from 100% post-consumer high-density polyethylene (HDPE) milk cartons. To arrive at this point, Invicta trialled and tested hundreds of recycled materials in the UK, Europe and the US, and created two new processes: rPETable and rNEWable. Greenpac were quoted in the article as saying that the processes make moulding in recycled materials very cost-effective and can lower carbon footprints significantly. Because these materials can be recycled repeatedly, they offer the additional benefit of further reducing the depletion of natural resources.

Another recent new development is the production of new food-grade PET resins by polymer suppliers that contain rPET material. For example, in 2012 it was reported in *European Plastics News* [26] and *PETplanet Insider* [27] that Artenius had developed a range of new PET resins that combine vPET and rPET, and that they had been approved for use in direct-food contact materials and articles. The new resins are called Artenius Unique and Artenius Elite, and are available in several grades (Chapter 8). Other resin suppliers are also launching food-grade rPET products. For example, in 2013, Phoenix Technologies announced LNO^{TM} w resin, a melt-extruded rPET pellet for ambient/cold/frozen food applications [28] that can be blended with vPET or used in food-packaging products at ≤100%.

3.5 Impact of bioplastics and oxobiodegradable additives

Over the past 15 years there has an increasing amount of interest in plastics that are not derived from fossil fuels, particularly for use in the food-packaging sector. These materials are referred to as 'bioplastics' and have several attractive properties, one of the major ones being sustainability because they are obtained from natural products. As mentioned in Section 3.4 in relation to the first 100% plant-derived PET bottle developed by PepsiCo [23], it is now possible to purchase a bio-PET manufactured from natural products such as sugar cane which is claimed to have virtually identical functionality to petroleum-based PET [29]. The appearance of these materials adds to an already complex market, and means that industry now has a choice of several PET materials:
- Conventional petroleum-based PET
- Bio-based PET

- Oxobiodegradable PET
- rPET

There are also other bio-based plastics that have been developed recently that could replace PET and affect its future and that of its recycling industry. An example of such a development was reported in *Bioplastics World* in 2013 [30]. In this article it was stated that a supplier of renewable chemicals, Alpla, had joined Avantium, a partner of Coca-Cola, in the development bio-based bottles. Avantium is responsible for YXY, a platform for catalytic technology to convert plant-based carbohydrates into chemical 'building blocks' for bioplastics such as polyethylene furanoate (PEF), bio-based chemicals or advanced biofuels. PEF bioplastic is 100% bio-based and recyclable and, according to Avantium, has properties superior to those of PET, including being a higher barrier to oxygen, carbon dioxide and water, which could extend the shelf-life of products and reduce production costs. With regard to recycling, preliminary tests have shown that the recycling of PEF could take place along very similar lines to those for PET [31].

If packaging manufacturers wish to use alternative materials to vPET or rPET, then another option is biodegradable, compostable bioplastics such as polylactic acid (PLA), which is derived from starch. The attractiveness of biodegradable plastics such as PLA is that they offer a way of dealing with the problem of waste plastics by biodegrading and decomposing under composting conditions. They have been certified as industrially compostable according to the standard EN 13432, and are particularly useful for single-use articles that could become mixed with food waste and bio-waste [32].

However, as the need to recover more post-consumer plastic increases due to environmental, economic and societal pressures, some of these alternative technologies are causing concern with respect to the integrity of recycling streams. Several problems may arise if biodegradable plastics and conventional amorphous plastics [e.g., polystyrene (PS) and acrylonitrile–butadiene–styrene] and semi-crystalline plastics (e.g., PET and Nylon) become mixed up during the collection of plastic waste. One of the main problems is that biodegradable plastics have, in general, lower heat stability and so degrade at the processing temperatures used for these conventional plastics. They can then discolour the conventional plastic and low-molecular weight (MW) compounds that can be formed can reduce their thermal stability. Polymers such as PLA in the plastic waste stream is a particular problem for PET because a relatively high processing temperature (>250 °C) is used due to its high crystalline melting point.

Other adverse effects for PET can result from this source of contamination. For example, a recent study carried out in Italy [33] showed that a small amount of PLA in a PET bottle processing line can significantly affect the rheological properties of the material. However, this study also found that the mechanical properties were affected only in certain instances, and that the thermal stability of the PET was not modified significantly. Another study by the Austrian Research Centre estimated that as little as 2% of biodegradable material in the recycling stream can have an influence on the quality of the recycled plastic. The presence of these types of plastics can, therefore, handicap the ability of the recycling industry to develop high-quality recyclates. In the UK, WRAP [34] have stated that introduction of compostable packaging can be successful in controlled environments, (e.g., London 2012 Olympic Games). In the same *Bioplastics World* article, it was highlighted that various venues use compostable foodservice products that can be collected after events and managed using local or acquired facilities for composting.

Another example of research carried out in this area is a collaborative project undertaken by the Materials Knowledge Transfer Network and Smithers Rapra to investigate the effect that a small amount of PLA entering a recycling stream can have on the physical properties (e.g., MW) of extruded PET [35]. The aim of the investigation was to determine if there was any change in viscosity and to establish if any changes that did occur were chemical or physical. To carry out the work, recovered PET from Closed Loop Recycling in Dagenham, UK, was dried at 130 °C and blended with extrusion-grade PLA in a twin-screw extruder. The results of the compounding trials revealed that when 1% of PLA was introduced to 99% PET, colour changes were visible, but no significant change in viscosity was observed. When the level of PLA reached 3%, the viscosity of the material was reduced to 82% of the original, and once the level of PLA reached 5%, it fell to 55%. These results showed that, at higher levels, PLA reduced the MW of the PET during extrusion by a chemical reaction.

Due to demand, the market for biodegradable packaging has grown over the last 5 years and is predicted to continue to grow at a steady rate for the rest of the decade. Data published by Smithers PIRA [36] show how strong this growth has been. The data presented in Table 3.4 are for biodegradable polymers (i.e.,

Table 3.4: Biodegradable packaging market by region – 2012 to 2018 (€ million).

	2012	2013	2014	2015	2016	2017	2018	Compound annual growth rate (%) 2008–2018
Europe	487	547	616	693	775	865	968	12.3
North America	467	531	603	685	771	866	974	13.3
Asia	550	615	688	770	866	974	1096	12.1
Rest of the World	37	59	99	162	168	187	220	7.6
Total	1,542	1,752	2,006	2,310	2,580	2,892	3,258	12.1

Note: Values were originally in USD but have been converted to euros at an exchange rate of $1 = €0.77

Reproduced with permission from M. Rao in The Future of Sustainable Packaging – Market Forecasts Until 2018, Smithers PIRA, Leatherhead, UK, 2013. ©2013, Smithers PIRA [36]

bioplastics such as PLA) as opposed to oxobiodegradable plastics. The growth rate of biodegradable packaging is strong, but it represents a small percentage of the overall packaging market. Use of biodegradable packaging is limited by its relatively high-cost, which can be 2–3-fold that of conventional, high-tonnage packaging plastics. This report also stated that use of PLA in particular is expected to expand between 2013 and 2018 owing to substantial capacity additions globally and the growing interest of packaging companies in sustainability.

Due to this increasing amount of biodegradable packaging in the market, even if effective steps are taken to segregate it from the conventional plastic waste stream, some will become mixed up with it. Hence, the ability to detect these types of contaminates will continue to be of commercial importance and, if effective segregation measures are not put in place, this capability could become essential. Such a detection system for the recycling of PET and HDPE, based on near-infrared spectroscopy, was developed during the EU-funded project SuperCleanQ (Section 6.5.2) [37]. In addition to being able to detect PLA in the polymer melt at 0.01%, the system could also detect other biodegradable contaminants that can find their way into recycling streams, oxobiodegradable additives (i.e., metal catalyst-based additives), and these could be detected in the PET melt at 0.1%. Oxobiodegradable additives are added to conventional thermoplastics such as PET and polyethylene (PE) to render them biodegradable. They ensure that under the influence of the ultraviolet light in sunlight the plastics decompose and fragment into very small pieces. The technology was developed to address such problems as plastic-bag litter in the environment. However, if these additives find their way into recycled polymers *via* contaminated recycling streams, they do not tend to influence processing stability. However, if they are present at sufficiently high concentrations, they can cause an undesirable reduction in the stability of any product manufactured from the recycled plastic due to catalytic breakdown upon exposure to sunlight (Section 7.6.2.2).

If there are concerns that this type of contamination is present in rPET it could discourage industry from using the material in their products, which would have a serious impact on the PET-recycling industry. Because of the possible impact on the recycling industry of the increasing use of biodegradable additives in PET waterbottles, the APR released a voluntary testing protocol in 2010 designed to examine the effect of such additives on recyclable PET [38]. This problem has resulted in research in development of in-line detection systems that can detect such additives (Section 7.6.2.2).

In another move to address some of these concerns over additives that impart biodegradability, Bioplastics International have marketed a specially formulated additive that makes polymers biodegradable without sacrificing the beneficial properties of the original polymer (including its ability to be recycled effectively). This pure organic additive is claimed to work with all plastics, including PET, PE, polypropylene, PS, polyvinyl chloride (PVC) and polycarbonate and will also work well with rubber products [39].

3.6 The future

The technology available for recycling PET has advanced continuously in recent years. For example, in 1999, it was reported that Erema had introduced a new recycling technology for PET that enabled rPET thermoformed sheets to be produced directly from post-consumer PET bottle flake [40]. This advance improved the efficiency with which the process could be carried out by eliminating the need for an interim step of pelletisation and the conventional steps of crystallisation and pre-drying. The Erema system included a cutter, compacter and extruder, with the bottle flakes being fed into the compacter where they are mixed, heated, crystallised and dried in one step. To reduce the loss of intrinsic viscosity and increase the evaporation rate the operation was done under a vacuum.

More recent innovations include development of a packaging film with ultra-high gas-barrier properties from a combination of PET and Plantic biodegradable film. The renewable and recyclable film (developed by Plantic Technologies and called Plantic eco Plastic R) was reported in an article in *Plastics and Rubber Asia* in 2013 [41]. The film contains 60% renewable materials and is fully recyclable, with the PET recovered in traditional recycling streams and Plantic's barrier material dissolving and biodegrading in the process.

Because of its attractiveness as a recyclable product, its cost and inherent properties, PET is being used to help replace existing products manufactured using materials or technologies that render them less recyclable (e.g., PET-based self-reinforcing polymer composites or all polymer composites to replace traditional glass fibre-reinforced plastics with good lightweight, mechanical and interfacial properties). PET is one of the most attractive polymers to be used in these fully recyclable all-polymer composites. At universities in London and Eindhoven [42], unidirectional all-PET composites have been prepared from skin-core structured bi-component PET multifilament yarns by a combined process of filament winding and hot-pressing. During the hot-pressing stage, the thermoplastic copolyester 'skin' or 'sheath' layers were melted selectively to weld high-strength polyester cores together creating an all-PET composite.

Another example is the change that Proctor & Gamble have made to the materials used in the manufacture of its Oral-B manual toothbrush [43]. The company has switched from PVC (which can be difficult to dispose of and recycle) to a combination of PET resin and Octal Petrochemical's proprietary direct PET (DPET™), which are both recyclable. Such developments provide new potential markets for PET, which would ensure the availability of PET for recycling in the future, as well as providing another potential re-use for rPET.

To assist with PET recycling, a company making biodegradable plastics, DaniMer, have recently developed a new bio-based label adhesive (designated 92721) for attaching labels to PET bottles [44]. The adhesive, which can also be used to attach labels to HDPE bottles, dissolves completely and unobtrusively in the PET flake caustic wash used at the start of the recycling process. This property has been shown by industry-standard protocol tests to provide zero levels of contamination to the PET recycle stream. The bio-based product is said to be priced competitively and its price is not linked to petroleum-derived substances.

Improving the robustness of PET so that it can withstand several successive recycling operations is another area receiving attention. The need for this feature has arisen because it is becoming possible to recycle post-consumer PET containers (e.g., bottles) back into new PET containers (i.e., 'cradle-to-cradle' recycling) and because the long-term effects that multiple recycles will have on the quality of the final container must be considered. Research is being carried out to improve the process capability of PET through the recycling stream. One of the potential solutions that look promising is improving the capability of the reheat technology in the processing of PET bottle preforms. This process can be carried out by incorporating a component with high absorption of infrared radiation into the material, which has also demonstrated that it can mask potential yellowing of PET due to multiple processing operations [45].

The recycling of plastic has participants (generators of recyclate and users) in all geographical regions. At the time of writing there is an overall movement of recycled plastic from Western countries (i.e., EU and US), where it is collected and separated, to Eastern countries (particularly South East Asia and China), where it is used in manufacturing. This situation exists due to the movement and concentration of manufacturing enterprises to this geographical region over the past 20 years due to the favourably low labour rates. However, this scenario could change due to increases in labour rates, the rising cost of transportation and potential legislation in areas such as the EU (Section 3.2) that might restrict the export of recovered plastic to retain resources and stimulate the indigenous recycling sector.

The situation that exists in 2016 for the recycling of PET (particularly for food contact applications) is encouraging, with a lot of research and development to achieve improvements in certain areas (e.g., sorting of black products and removal of barrier layers from post-consumer products) ongoing, and many 'super-clean' processes being considered by the European Food Safety Authority in the EU and the FDA in the US (Chapters 4 and 6). Interest in recycling PET is increasing throughout the world, and the amount of material collected and re-used for food and non-food uses is expected to show strong growth for many years to come.

References

1.   F. Welle, *Resources, Conservation and Recycling*, 2013, **73**, 41.
2.   V. Goodship in *Introduction to Plastics Recycling*, 2nd Edition, Smithers Rapra, Shawbury, UK, 2007.
3.   *Plastics Packaging – Recycling by Design*, Revised Edition, Recycling Of Used Plastics (RECOUP) Ltd, Peterborough, UK, 2009.
4.   Anon, *Plastic Packaging Innovation News*, 2007, **2**, 22, 1.

5.   Anon, *Plastic Packaging Innovation News*, 2006, **2**, 9, 1.
6.   http://www.petcore.org
7.   http://www.plasticsrecyclers.eu
8.   F. Welle, *Resources, Conservation and Recycling*, 2011, **55**, 865.
9.   http://www.napcor.com
10. http://www.letsrecycle.com/news/latest-news/plastic-exports-jeopardise-domestic-recycling-petcore-warns/
11. http://www.plasteurope.com/news/PET_RECYCLING_EUROPE_t226075/
12. http://www.pcipetpacking.com
13. Anon, *PETplanet Insider*, 2011, **12**, 10, 16.
14. *Towards a Circular Economy: A Zero Waste Programme for Europe*, COM(2014) 398 Final, European Commission, Brussels, Belgium, 2nd July 2014.
15. http://www.epa.org
16. *Report on Postconsumer PET Container Recycling Activity in 2014*, NAPCOR/ APR,Napor, 13th October 2015. http://www.napcor.com/PET/pet_reports.html
17. Anon, *SA Plastics Composites and Rubber*, 2012, **10**, 2, 13.
18. T. Gerstl, *PETplanet Insider*, 2012, **13**, 7–8, 27.
19. R. Maiseviciute in *PET Packaging – Global and European Trends and Opportunities*, Euromonitor International, London, UK, 7th November 2011.
20. Anon, *British Plastics and Rubber*, 2001, April, 21.
21. Anon, *Plastics and Rubber Asia*, 2012, **27**, 187, 18.
22. Anon, *European Plastics News*, 2011, **38**, 9, 34.
23. Anon, *Chemical Weekly*, 2011, **56**, 34, 142.
24. Anon, *PETplanet Insider*, 2012, **13**, 1–2, 26.
25. Anon, *British Plastics and Rubber*, 2013, March, 9.
26. Anon, *European Plastics News*, 2012, **39**, 9, 42.
27. Anon, *PETplanet Insider*, 2012, **13**, 7–8, 31.
28. Anon, *PETplanet Insider*, 2013, **14**, 6, 24.
29. G. Crosse, *Bioplastics World*, 2012, **2**, 5, 5.
30. Anon, *Bioplastics World*, 2013, **1**, 11, 4.
31. P. Mangnus, *Bioplastics Magazine*, 2012, **7**, 4, 12.
32. K.B. Lange, *Materials World*, June 2014, p.24.
33. F.P. La Mantia, L. Botta, M. Morreale and R. Scaffaro, *Polymer Degradation and Stability*, 2012, **97**, 1, 21.
34. Anon, *Bioplastics World*, 2013, **1**, 1, 5.
35. L. Asfa-Wossen, *Materials World*, 2010, **18**, 12, 4.
36. M. Rao in *The Future of Sustainable Packaging – Market Forecasts Until 2018*, Smithers PIRA, Leatherhead, UK, 2013.
37. SuperCleanQ EU-funded FP7 Research Project. http://www.supercleanq.eu
38. Anon, *Additives for Polymers*, 2010, March, 4.
39. A Buan, *Plastics and Rubber Asia*, 2014, **29**, 203, 17.
40. Anon, *Asian Plastics News*, 1999, September, 57.
41. Anon, *Plastics and Rubber Asia*, 2013, **28**, 198, 8.
42. J.M. Zhang and T. Peijs, *Composites Part A*, 2010, **41**, 8, 964.
43. Anon, *Plastics and Rubber Asia*, 2012, **27**, 188, 22.
44. Anon, *International Bottler and Packer*, 2012, **86**, 9, 16.
45. Anon, *Macplas International*, 2010, March–April, 12.

4 Regulations and guidance documents from the european union and US food and drug administration

4.1 Introduction

This section in the book provides a brief introduction to two of the principal types of regulations most applicable to the recycling of polyethylene terephthalate (PET):

- Regulations that have been published to provide safeguards for the environment and to encourage recycling and achievement of recycling targets.
- Regulations that apply to food-contact materials and articles made from virgin plastics and recycled plastics.

It is important to provide information on such areas because they have two crucial roles. First, they help to ensure that the starting material for the PET-recycling industry (i.e., post-consumer PET and waste PET) is available in sufficient quantities to make the industry viable. Second, they help to achieve market and consumer acceptance of suitably prepared recycled polyethylene terephthalate (rPET) for one of its biggest markets: food packaging.

As mentioned in Chapters 1 and 2, the interest and commercial activity associated with the recycling of PET and other plastics has arisen due to a general global trend towards an increasing desire to improve sustainability in all areas of life. To help focus and address this desire, particular policies have been pursued in several countries and geographical areas. A prominent example of such a policy is the Integrated Product Policy (IPP) [1] in the European Union (EU), which has been in existence for ≈20 years. In addition to these policies, several directives and regulations have been published in the EU to help achieve environmental and sustainability goals [e.g., the Landfill Directive (1999/31/EC) and the Packaging and Packaging Waste Directive (94/62/EC)]. The approach taken in the US is much less centralised, and publication of measures to protect the environment tend to be taken at state level rather than at federal level.

Food contact regulations are the other main type of regulation covered in this section because a large amount of the PET collected for recycling is food-grade material and, once it has been decontaminated using one of the processes approved by the European Food Safety Authority (EFSA) or US Food and Drug Administration (FDA) described in Chapter 6, it can be used for food contact products. Hence, the regulations for food contact that exist in different countries (particularly the EU and the US) are of particular importance to rPET. In contrast to the situation that exists with recycling regulations, the EU and US have centralised regulatory systems, though in the EU these 'harmonised' regulations (as they are referred to) have not been extended to all food

https://doi.org/10.1515/9783110640304-004

contact materials. However, plastics, which as a class includes PET, are covered by a harmonised regulation, the Plastics Regulation (EU) 10/2011, and there is also a specific EU regulation that covers use of recycled plastic for the manufacture of material and articles intended for food contact applications, (EC) 282/2008.

Food contact regulations have been introduced in most of the developed nations and regions in the world. These systems can vary considerably in detail, but they usually operate on at least three levels, as described below in Sections 4.1.1–4.1.3.

4.1.1 General compositional requirements

General compositional requirements are addressed by ingredients that may be used in the manufacture of a product and is often referred to as a 'Positive List'. The ingredients on this list will have an established toxicity profile put together by an organisation such as EFSA, which will have established that they can be used in this type of application, but also whether specific migration limits are required to restrict how much can migrate into food. There may also be a 'Negative List', which states which substances must not be used, but these are rarer because the list is potentially infinite and tends to be restricted to substances of high concern within a particular class of product [e.g., initiators for ultraviolet (UV)-curable inks].

4.1.2 Tests to show compliance with migration and specific compositional requirements

Tests to show compliance with migration and specific compositional requirements can take two main forms:
- Food-migration tests using food simulants or food products that cover overall (also called 'global') migration and specific migration of substances that have restrictions [i.e., specific migration limit (SML)] in the regulations.
- Compositional tests to quantify specified substances that have restrictions in the regulations, such as residual monomers, heavy metals, primary aromatic amines and breakdown and/or reaction products from certain additives (e.g., antidegradants) or the polymer itself.

4.1.3 Good manufacturing practice

Good manufacturing practice ensures that once the material or article has complied with the compositional and migratory aspects of the regulations, it continues to be manufactured within a quality system to ensure that changes to formulation, and manufacturing parameters do not take place in an uncontrolled manner, and that the substances used in the manufacture of the product (and its environment of manufacture) are of a high purity and high standard to ensure that it is not contaminated.

In the case of using rPET to produce non-food products (e.g., fibres or strapping), those products will have to be manufactured in such a way that they can achieve the criteria for quality and performance laid down by the international, national or industry standards relevant for the particular PET product in question. Some of the tests that can be carried out on rPET products to achieve this compliance are described in Chapter 7.

In addition to the EU and US, other developed countries and regions in the world have put in place a framework of regulations to help achieve higher rates of recycling and recovery and reduce the impact of their economies on the environment and to protect their consumers from food contact materials and articles that represent a hazard to health. These are also mentioned briefly in this section, but readers looking for a more thorough treatment can find it in publications such as the one edited by Rijk and Veraart [2–5].

4.2 Regulations and guidelines set by the European Union

4.2.1 Introduction

Regulations and directives are the way in which laws are introduced in the EU. One of the principal differences between these two types of instrument is that regulations become law throughout the EU once they have been published, whereas directives must be incorporated into the law of a nation state before they become law. Both usually have three main sections: recitals, articles and annexes:

1. *Recitals* are policy statements providing the background to the regulation and other information (e.g., related regulations, committees involved), parameters taken into account in drafting the regulation.
2. *Articles* are the laws that must be understood and obeyed.
3. *Annexes** can be regarded as the 'working part' of the regulation which describe 'how to comply' with the laws in the Articles.
   *The number of annexes in a regulation or directive can vary and the more complex regulations will have several them. For example, the EU Plastics Regulation (EU) 10/2011 has six annexes and some are extensive (e.g., Annex I, the 'Union List' of permitted substances is 60 pages long).

PET is used extensively for food packaging, particularly for bottles and thermoformed products (e.g., trays). This PET packaging waste is collected and recycled for re-use in food and non-food applications (Chapters 8 and 9). In the case of its use for food contact materials and articles, there are several specific EU regulations that address this sector and these are covered below in Sections 4.2.3.

4.2.2 European Union regulations and policies for recycling

The EU has been publishing regulations to manage waste for many years. For example, in the mid-1980s it introduced the first measure on the management of packaging materials. This was Directive 85/339/EEC and it set rules for several areas, including the production, use and recycling of packaging. Specifically, it covered containers that contained liquid for human consumption (e.g., bottles) and the disposal of such containers. Since the mid-1980s, the environmental concerns referred to in Chapters 1 and 3, and the massive increase in the use of plastic packaging and domestic and commercial waste in general, has resulted in a dramatic increase in the amount of environmental legislation. One such piece of legislation, which goes back to 1994, is the Packaging and Packaging Waste Directive (94/62/EC). This Directive harmonised national measures on the management of packaging and packaging waste and aimed to:

– Avoid obstacles to trade in the EU and the restriction of competition.
– Provide a high degree of environmental protection by setting targets for the presence of certain heavy metals in packaging [i.e., cadmium, mercury, lead and chromium(VI)].
– Promote recycling by the setting of targets for recovery and recycling.

Other important directives were to follow. In 1999, the Landfill Directive (1999/31/EC) was published and in 2008 the revised Waste Framework Directive (2008/98/EC) was published. The former covered all the different generic classes of material that could be placed into landfill sites. It also stated particular materials that were to be excluded from landfill sites (e.g., quarry waste and waste tyre-derived materials) and was designed to supplement the Waste Framework Directive by:

– Reducing the negative effects of landfilling.
– Providing uniform technical standards for landfills.
– Stating particular requirements for landfills (e.g., locations and monitoring protocols).

One other regulation was the Integrated Pollution Prevention and Control (IPPC) Directive (2008/1/EC), which required that all non- inert landfilling sites, apart from the very smallest, had to be regulated and that sites above a certain capacity needed a permit to continue. The Landfill Directive can also be regarded as supplementing the IPPC Directive by setting safety standards for the operation of landfill sites.

When focussing on the activities surrounding specific products, the IPP concept, which promotes good environmental practice throughout the lifecycle of a product (Section 4.1), has been in existence in Europe for ≈20 years. The IPP seeks to promote good environmental practice throughout the lifecycle of all manufactured products or services and uses protocols for lifecycle assessment to assess if this has been achieved. In the case of manufactured products, manufacturers are required to consider the entire lifecycle of their products (i.e., design, production, use in service and disposal phases).

The key points of the IPP are:
- Managing wastes
- Production innovation
- Market creation
- Transmission of information
- Allocation of responsibility (e.g., for disposal)

Because of the very large range of products in society, these points have been addressed by various tools:
- Economic measures
- Substance bans
- Environmental labelling
- Guidelines for product design

One of the mechanisms by which these goals have been advanced is the introduction of EU directives and regulations. For example, three EU directives that can be said to contain IPP ideology are the Waste Electrical and Electronic Equipment Directive (2002/96/EC), End-of-Life Vehicle Directive (2000/53/EC) and the Packaging and Packaging Waste Directive (94/62/EC).

The Packaging and Waste Packaging Directive (94/62/EC), one of the earliest pieces of legislation and, as referred to above, had a major impact on the recycling of plastics in general, and PET in particular, because the packaging sector is the principal consumer of plastics in the EU, taking >40% of the total market (Table 2.1). This Directive covered plastics and other types of material (e.g., paper and board) used to produce packaging products, and it sought to harmonise the management of packaging waste to protect the environment. Three important aspects of this legislation were: (i) the introduction of restrictions for mercury, lead, cadmium and hexavalent chromium in packaging; (ii) targets to be achieved by 2008 for plastic recycling (≥22.5% was set); and (iii) energy recovery should rise to 50% of that recycled.

4.2.3 Regulations that cover food contact materials set by the European Union

In addition to specific regulations that address food contact plastics (Section 4.2.4), there are regulations that cover all food contact materials within the EU:
- Plastics, rubbers, thermoplastic rubber, adhesives, coatings, and printing inks
- Non-polymers (e.g., ceramic, wood, cork, metals, paper and board)

Two regulations fall into this category: the Framework Regulation (EC) 1935/2004 and the Good Manufacturing Practice Regulation (EC) 2023/2006.

4.2.3.1 Framework Regulation (EC) 1935/2004

In addition to applying to all food contact materials, Framework Regulation (EC) 1935/2004 also applies to all food contact products, for example:
– All elements of food packaging
– Cookware, cutlery and tableware
– Work surfaces
– Food contact parts of processing equipment

It also applies to articles that can reasonably be expected to be brought into contact with food, (e.g., linings inside refrigerators).

As the name suggests, this measure provides the framework for EU regulations and also contains important articles such as Article 3, which stipulates that food contact materials and articles:
– Shall be manufactured in compliance with good manufacturing practice (GMP) (2023/2006/EC)
– Do not transfer constituents to food in quantities which could:
    – Endanger human health; or
    – Bring about unacceptable change in composition of food; or
    – Bring about deterioration in organoleptic properties.
– Labelling, advertising and presentation shall not mislead the consumers

To demonstrate compliance with the Framework Regulation it is necessary to:
– Carry out organoleptic testing on the finished article using standard methods*
– Show that the composition of materials or articles complies with the following in any applicable EU regulation or national regulation:
– Positive list
– Overall migration limit
– Any applicable specific migration limits
– Any applicable residual limits
– Any applicable specific test requirements (e.g., metals)
    *For example, taint and odour according to BS EN 1230:2009 Parts 1 and 2 and colour migration according to the test method provided by the Council of Europe Resolution on Inks AP(89)1. For the taint and odour tests, an appropriate food product is used (e.g., chocolate) and, for the colour-migration test, food simulants such as distilled water, 3% acetic acid, 15% ethanol or olive oil can be used.

4.2.3.2 The good manufacturing Practice Regulation (EC) 2023/2006

As shown above, the GMP regulation is tied to the Framework Regulation by being referenced within it. It ensures that, in addition to meeting all of the compositional and testing requirements, food contact materials and articles are also manufactured correctly and traceably to ensure that ingredients of the correct purity are used and added to the material in the appropriate manner, and that no contamination is picked up at any stage during manufacturing, storage and packing.

In addition to applying to all of the classes of food contact material listed in Annex 1 of the Framework Regulation, the GMP regulation also applies to:
- Combinations of those materials
- Recycled versions of those materials

The GMP regulation also applies to *all sectors and stages* of manufacture, processing and distribution of materials and articles (e.g., plastic resins, additives, masterbatches, final products).

4.2.4 Regulations set by the European Union that apply to plastics

The EU trading block is undergoing a lengthy and extensive 'harmonisation' programme in which national regulations are being replaced with EU-wide regulations. This programme has been underway for >20 years, but it has been completed only for a small number of food contact materials. Plastics and recycled plastics are two of these materials, and the others are:
- Regenerated cellulose
- Ceramics
- Active and intelligent materials

For all the other classes of food contact material listed in Annex I of the EU Framework Regulation (EC) 1935/2004 (Section 4.2.3), harmonised EU regulations are absent. Hence, the applicable national regulations (e.g., German BfR Recommendations) have to be used to determine if a product is safe to use with food (Section 4.2.5).

4.2.4.1 European Union plastics regulation (EU) 10/2011
The Plastics Regulation (EU) 10/2011 was published on 14th January 2011 and brought to a conclusion a long programme to harmonise EU regulations for plastics. It is a single-source reference for all plastics and multiple-material products that contain a plastic layer (though it has been amended several times and these amendments have yet to be consolidated into it). When it came into force on 1st May 2011, it repealed the Plastics Directive 2002/72/EC and the two vinyl chloride Directives: 80/766/EEC and 81/432/EEC.

The regulation is substantial and contains:
- An overall migration limit of 10 mg/dm^2
- Positive list (called the 'Union List') of ≈900 starting substances and additives

- Specific migration limits for certain substances
- Restrictions for certain classes of substances (e.g., heavy metals)
- Protocols for overall migration testing and specific migration testing

With respect to the migration testing of plastics, this is covered in the Annexes, as shown below.

The food simulants that should be used for work on overall migration and specific migration are specified in Table 1 of Annex III (Table 4.1).

Table 4.1: Food simulants for overall and specific migration work.

Food simulant designation	Composition
Food simulant A	10% *(v/v)* ethanol
Food simulant B	3% *(w/v)* acetic acid
Food simulant C	20% *(v/v)* ethanol
Food simulant D1	50% *(v/v)* ethanol
Food simulant D2	Vegetable oil*
Food simulant E	Poly(2,6-diphenylene-*p*-phenylene oxide); particle size, 60–80 mesh; pore size, 200 nm

* Any vegetable oil with a distribution of fatty acids as defined in Annex III

Table 2 in Annex III provides a list of food types/food simulant matches to assist with the selection of the appropriate food simulant to meet the end-use requirements of a particular product. The foods listed in this table are real food products (e.g., butter, dry pasta or chocolate) or well-defined categories of food [e.g., milk and milk- based drinks (whole, partly dried, skimmed or partly skimmed)].

Having selected the appropriate food simulant(s) for a food contact product from the information supplied in Annex III, guidance is required regarding the protocol to be used for migration testing, and this is provided in Annex V. Annex V provides tables, which can be used to determine which conditions (i.e., temperature and time) should be used for the migration tests to reflect the end-use conditions of a particular product. There are two tables in Chapter 2 of this Annex, which deal with the contact time and contact temperature that should be used for specific migration work. Overall migration work is covered in Chapter 3, in which a table provides seven standardised testing conditions (OM1 to OM7) that can be used depending upon the intended food contact conditions of the product to be tested. Also present in Annex V is advice for dealing with contact conditions not covered specifically by these tables, and other information that can be of use when carrying out these types of tests (e.g., correction factors, such as the reduction factor for fatty foods).

The approach that would be applied for a product made from food- grade rPET would, therefore, be to determine the types of food that the product might realistically be expected to contact during its normal lifetime and use this information to obtain the food simulant (or mixture of food simulants) that should be used for the migration testing from the information provided in Tables 1 and 2 of Annex III. Then, after consideration of the expected range of contact times and contact temperatures that the product would contact these food types, use the table in Chapter 3 of Annex V to determine which contact time and contact temperature to use for the overall migration testing. If the rPET product contains substances with specific migration limits, then the tables in Chapter 2 of Annex V can be used to determine the contact time and contact temperature for this work, and then the analysis methods provided in EN 13130 (Section 7.4.2) can be used to quantify the specific migrants. Other points that would need to be considered include if the migration testing (overall and specific) be carried out using a total-immersion approach (whereby both sides of the product contact the food simulant) or if a single-sided approach [if only one side of the product contacts the food in service (e.g., a food tray)] should be used.

Taking the example of a PET food tray manufactured using food- grade rPET that could be used for a range of oven-cooked ready meals, the testing regimen that could be used on the final article is:
- Food contact product: rPET food tray
- Food types: fatty, acidic and aqueous food types
- Contact time and contact temperature: ≤1 h at 200 °C
- Selection of food simulant: 10% ethanol, 3% acetic acid and olive oil

Testing conditions for the overall migration work:
- OM6 (4-h reflux) for 10% ethanol and 3% acetic acid
- OM7 (2 h at 175 °C) for olive oil

Testing conditions for the specific migration work:
- 4-h reflux for 10% ethanol and 3% acetic acid
- 1 h at 200 °C for olive oil

The specific migrants that would be targeted by the specific migration work (because they have SML values in the Union List in (EU) 10/2011) are: ethylene glycol (EG) (SML = 30 mg/kg), diethylene glycol (DEG) (SML = 30 mg/kg) and terephthalic acid (TPA) (SML = 7.5 mg/kg). The test methods to be used for these determinations are found in EN 13130 – Part 2 (TPA) and Part 7 (EG and DEG).

To complete the testing programme, examples of the final product must pass the organoleptic tests to completely satisfy the requirements of Article 3 in the EU Framework Regulation (Section 4.2.3.1).

Once the regulatory assessment of any plastic product has been completed, the information obtained must be summarised in a way that can be understood easily by the marketplace. For this purpose, (EU) 10/2011 assists companies and independent testing bodies by having an annex (Annex IV) that provides guidance as to what should be included in a compliance statement

4.2.4.2 European Union recycled Plastics Regulation (EC) 282/2008

European Union Recycled Plastics Regulation (EC) 282/2008 applies to all plastics recycled in which the intention is using the recycled plastic for food contact materials and articles. At present, recycling is not widespread in the EU for all food contact plastics, but it is at an advanced stage for PET (particularly PET bottles). This regulation also covers articles and materials in the Plastics Regulation (EU) 10/2011 that contain some recycled plastic, and stipulates that the plastic must be sourced from a recycling process that meets the definitions in Regulation (EC) 282/2008, and that they are manufactured according to the Good Manufacturing Practice Regulation (EC) 2023/2006.

With regard to the definitions mentioned above, Regulation (EC) 282/2008 does not apply to:
- Plastics made with monomers and starting substances derived from the depolymerisation of plastic materials.
- Plastics made from in-process scrap that complies with (EU) 10/2011.
- Recycled plastic used behind a functional barrier as defined in (EU) 10/2011.

Regulation (EC) 282/2008 describes 'challenge' tests, which are used to determine if a particular recycling process for plastics is effective at removing contaminants from the waste feedstock. These challenge tests involve 'doping' the process using a series of chemical compounds that are representative (in terms of molecular weight, polarity, and volatility) of the types of compounds that could contaminate post-consumer waste. The regulation also specifies the procedure for applying to the European Food Safety Authority (EFSA) to obtain a 'scientific opinion' for a recycling process. EFSA maintain a register of valid applications for the authorisation of recycling processes on their website [6]. The information presented in this register includes the type of polymer that is being recycled for food use (most applications at present target PET), the name of the business operator intended to be the authorisation holder, and the type of process being used. With regard to assessment of these applications, EFSA adopted its first scientific opinions on the safety of specific recycling processes in 2012 [7]. Since then, there has been a steady stream of opinions released by the EFSA. The EFSA have published guidelines on the criteria to be used for the safety evaluation of a mechanical recycling process to produce rPET for food use (Section 4.2.6).

In summary, a rPET food contact must be capable of passing the requirements of the Framework Regulation (EC) 1935/2004, the GMP Regulation (EC) 2023/2006, the compositional requirements, migration criteria and applicable restrictions laid down in (EU) 10/2011, and it must have been recycled using a recycling process that has met the criteria described in (EC) 282/2008 to ensure that it has been de-contaminated to a sufficient degree. If used as a packaging material, it must also comply with EU Directive 94/62/ EC with respect to the restrictions for lead, cadmium, mercury and hexavalent chromium.

4.2.5 National regulations for food contact in the European Union

National Regulations for Food Contact in the European Union are the regulations that existed in Europe before the EU started to create its own regulations and harmonise some of them (e.g., those for plastics).

Seven EU Member States (including the UK, Germany, France, Italy and The Netherlands) have some form of regulation for food contact materials. The national regulations in each country establish:
– Means of enforcement
– Offences for failing to comply

The EU *Practical Guide for Testing Food Contact Materials* [8] states 'if no specific EU Regulation exists for a material – refer to relevant National legislation' and the German BfR Regulations are often used in this instance because they are comprehensive and accepted widely within Europe.

The German BfR Regulation that covers PET is Recommendation XVII and any PET, including rPET, used in food contact materials and articles must comply with this Recommendation. In 2000, the German BfR issued a statement to ensure the safe mechanical recycling of plastics made from PET for the manufacture of articles for direct food contact. This statement has been incorporated into Recommendation XVII and introduces two criteria:
– An analytical assurance that post-consumer rPET must not be disadvantageously distinguishable from virgin polyethylene terephthalate (vPET).
– A migration limit of 10 ppb for evaluation of the performance of 'super-clean' recycling processes – but this limit is not to be used as a toxicological-based end-parameter.

4.2.6 Guidance documents for recycling food contact plastics in the European Union

Following on from the publication of EU Regulation (EC) 282/2008 (Section 4.2.4.2), in 2008 the EFSA published [9] an important guidance document: *Guidelines on submission of a dossier for safety evaluation by EFSA of a recycling process to produce recycled plastics intended to be used for manufacture of materials and articles in*

contact with food. These guidelines were adopted on 21 May 2008 after public consultation and discussion in panel. It contained information that covered areas, such as:
- How to submit an application to the EFSA.
- The information to be included in the application (e.g., contents of the Technical Dossier).
- Under which conditions a re-evaluation of a process could be requested by the EFSA.
- Information on the application of a quality system to a process.

Within the Technical Dossier section, these EFSA guidelines address determination of the contamination efficiency of a recycling process and provide guidance of the choice of surrogate compounds to use for the challenge test by referring readers to published sources that have proposed sets of surrogates for particular plastics. In the case of PET, the references provided include a paper published by Begley and co-workers [10], an EU guidance document by Franz and co-workers [11], and a FDA guidance document featured in Section 4.3.4. The recommendations, guidelines and criteria present in the report by Franz and co-workers can be considered as implementation of the results and conclusions from the European project FAIR-CT98-4318 [12].

Another important EFSA guidance document applicable to PET was published in the *EFSA Journal* in 2011 [13]: '*Scientific Opinion on the Criteria to be used for Safety Evaluation of a Mechanical Recycling Process to Produce rPET Intended to be used for Manufacture of Materials and Articles in Contact with Food'*. It was published by the EFSA Panel on food contact materials, enzymes, flavourings and processing aids. The purpose of the EFSA in publishing these guidelines was evaluation of submissions for PET recycling processes, and they developed criteria specific to that material. The EFSA Panel thought these criteria should be published for the sake of transparency and to inform all stakeholders on the considerations followed for the risk assessment of PET recycling processes. The scientific opinion, therefore, described the risk-assessment approach used by the EFSA Panel and provided arithmetic opinion values for the criteria specific to the evaluation of recycling processes for PET intended to be used in food contact materials, and covered several topics:
- Overview of contamination data on post-consumer PET bottles
- Data on sorption of chemicals into PET
- The challenge test: decontamination/cleaning efficiency of the recycling process
- Criterion of migration of potential contaminants
- Application of the key parameters for the evaluation scheme

With regard to non-food PET packaging in the recycle waste stream, the document states that the EFSA Panel considered it appropriate 'that the proportion of PET from non-food consumer applications should be no more than 5% in the input to be recycled'.

Guidelines set by the International Life Sciences Institute (ILSI) for recycling food contact plastics were published in 1998 [14], and share many of the same scientific principles established by the FDA (Section 4.3.4), but have some differences developed from new scientific knowledge and differences in approach to the same problems.

These guidelines were developed initially for food-grade plastics and use the concept of the challenge test conducted using surrogate chemical species. These species are listed in Table 4.2 and are similar to those used by the FDA (Section 4.3.4 and Table 4.3).

Table 4.2: Surrogate chemical species used in challenge tests.

Compound class	Substance	Concentration (%)
Polar, volatile	Trichloroethane	1
Polar, non-volatile	Benzophenone	1
Non-polar, volatile	Toluene	10
Non-polar, non-volatile	Chlorobenzene	1
Non-polar, non-volatile	Phenylcyclohexane	1
Polar, non-volatile (derived from organometallic)	Methyl palmitate (or methyl stearate)	1

Table 4.3: Surrogate compounds suggested by the FDA for a challenge test.

FDA category	Suggested compounds	Concentration in cocktail
Volatile polar	Chloroform Chlorobenzene 1,1,1-Trichloroethane Diethyl ketone	10 v/v
Volatile non- polar	Toluene	10% v/v
Non-volatile polar	Benzophenone Methyl salicylate	1% v/v
Non-volatile non-polar	Tetracosane Lindane Methyl stearate Phenylcyclohexane 1-Phenyldecane 2,4,6-Trichloroanisole	1% w/w
Heavy metal	Copper(II) 2-ethylhexanoate	1% w/w
Balance compounds for cocktail	2-Propanol as a solvent for copper(II) 2-ethyl hexanoate Hexane or Heptane as a solvent for the overall cocktail	10% v/v 68% v/v

v/v: Volume of surrogate per unit volume of entire cocktail
w/w: Mass of surrogate per unit mass of entire cocktail

The report on PET resulting from the large European FAIR- CT98-4318 'Recyclability' project [12] recommended the same surrogates as those shown in Table 4.2 but without the use of trichloroethane. As mentioned above, the results from the FAIR- CT98-4318 project were used to produce the EU guidance document for PET [11].

The key difference in the EU approach from the FDA protocol is the way the surrogates are mixed with PET. The FDA uses a solvent-based 'cocktail' whereas the ILSI approach uses direct mixing and sorption by mixing at 50 °C for 7 days, which was found to be functionally equivalent but simpler to prepare. The level of surrogates in PET flake should be 50–350 ppm for all except toluene, which should approach 500 ppm. Then, the contaminated flake is directed straight into the systems that first undergo the washing and drying steps, before 'super-cleaning' takes place.

The normal requirements of the challenge test are that 100% of the test articles are contaminated with surrogates and the results from the cleaning process must demonstrate 'not detectable migration' at the limit of detection (10 µg/kg) of the analytical methodology.

4.3 US regulations and guidelines

4.3.1 Introduction

The US does not have a national recycling law, and waste collection in the US is the responsibility of state and local governments. What the US does have, though, is a national law for management of solid waste called the Resource Conservation and Recovery Act (RCRA), which was passed by Congress in 1976, and abolished open dumps and required the US Environmental Protection Agency (EPA) to write regulations for hazardous waste management and guidelines for disposal of solid waste. It was the RCRA (which remains the cornerstone of federal legislation for recycling of solid waste) which made the state and local governments responsible for the collection and disposal of solid waste [15].

Due to the absence of comprehensive federal legislation, recycling legislation in the US is dealt with at state level or local (e.g., city) level. For example, several states and local jurisdictions have adopted laws and regulations affecting the production, use, and disposal of packaging. These provisions include standards for environmentally acceptable packaging (e.g., Coalition of Northeastern Governors heavy metal limits), minimum requirements for recycled content, restrictions on the use of certain substances in packaging (e.g., as described in California Proposition 65), and bans on particular types of plastic packaging.

Areas in which the federal government has legislated include:

- Regulating packaging for food, drugs and cosmetics (Section 4.3.3).
- Promoting government procurement of recycled products.
- Issuing guidelines to ensure manufacturers do not make incorrect claims about the environmental benefits of their packaging.

One of the main considerations that led to the growth of these packaging laws in the US during the 1980s and 1990s was the concern about the effect on the environment of disposing of post-consumer packaging. This concern was reflected in other areas of the world, but was particularly strong in the US because it landfilled >80% of its solid waste throughout the twentieth century [16].

A system that has proved of great use in the promotion of recycling plastics in general in the US and in other areas of the world (e.g., the EU) was developed by the Society of the Plastics Industry (SPI) in 1988, but has been administered by the American Society for Testing and Materials International since 2008. The SPI resin-identification code contains a number surrounded by 'chasing arrows' and is followed by an abbreviation for the specific plastic that it represents. The code has been adopted by legislation in most states in the US and the SPI resin-identification code for PET is '1'. This code appears on plastic packaging and is very useful in enabling users and recyclers to determine visually which plastic the product had been made out of.

4.3.2 Recycling regulations in the US

On a national level in the US the EPA, which encourages people to 'reduce, reuse, recycle' on its website [17], oversees various recycling issues, including regulation of hazardous wastes, landfill regulations and setting recycling goals. The EPA can publish rules that apply directly or indirectly to recycled plastics. One of these is *The Identification of Non-Hazardous Secondary Materials That Are Solid Waste* rule, which was published in 2012 and determines whether non-hazardous secondary materials (e.g., plastics) are considered 'fuels' or 'waste' when burnt. Combustion units that burn non-hazardous secondary materials classified as fuels are regulated as boilers under Section 112 of the Clean Air Act, whereas units that burn materials classified as wastes are regulated as incinerators under Section 129 of the Clean Air Act.

As mentioned already, more specific recycling legislation is localised through state or city governments. State regulation falls into two major categories: landfill bans and recycling goals. Landfill bans, which are in place in states such as Wisconsin, Michigan and North Carolina, make it illegal to place specified items in a landfill site, and these items can include recycled plastics collected in curbside recycling programmes. States which focus on recycling targets instead of landfill bans include California and Illinois. In some cities in the US (e.g., Seattle) mandatory recycling laws have been created that may fine citizens who throw away certain recyclable materials.

4.3.3 Food contact regulations in the US

Food contact regulations in the US are to be found in the Code of Federal Regulations (CFR). The CFR consists of over 40 titles, which are broken down into chapters (1, 2, 3...), sub-chapters (e.g., A, B, C...) and parts (1, 2, 3...).

CFR Title 21 is the important title in the CFR for food contact regulations and these appear in Sub-chapter B, within Parts 170 to 190 [18].

Different parts within 170 to 190 deal with different classes of additives, polymers and materials. One of the ways in which the FDA regulations differ from the EU regulations is that specific materials (e.g., olefin polymers) have their stand-alone sections within CFR Title 21 with the grades of resin that are compliant, a list of additives and, sometimes, migration tests. In contrast, the EU regulations tend to focus on generic classes of material (e.g., plastic), do not specify specific grades, and have an approved list of monomers and additives (the 'Union' list) for their manufacture.

Some of the classes of materials that are to be found in CFR Title 21, Parts 170 to 190, with their parts numbers are:
- Part 175: Adhesives and coatings
- Part 176: Paper and paperboard
- Part 177: Polymers

Part 177 contains sub-parts for different classes of polymers, for example:
- Part 177.1500 – Nylon resins
- Part 177.1520 – Olefin polymers
- Part 177.1630 – PET

With respect to the sub-parts that deal with specific polymer groups, these can contain the following:
- List of approved monomers.
- Specification for the polymer (e.g., density, melting point range, total extractables, intrinsic viscosity and softening point).
- Test methods to use for the tests to confirm compliance with the specification.
- Food-migration tests that must be carried out on the final product or test pieces.

With regard to additives that can be classified as colourants and pigments, there is no single specific section in CFR Title 21 to reference. There is a section for polymers, which is titled 'Colorants for Polymers' (178.3297), but it is also possible to use colour additives that are listed in CFR Title 21 for direct use with food (found in Parts 73, 74, 81 and 82), or Paper and Board (176.170), Resinous and polymeric coatings (175.300), or other appropriate sections, providing any 'restrictions' that are cited are obeyed.

To be compliant with the FDA regulations, a PET food contact product has to pass the following assessment:

1. *PET base resin*: The grade of PET used in the production of the product must meet the requirements stated in Part 177.1630 'PET polymers'.
2. *Pigments in the PET product*: Any pigments or colourants used in the PET must meet the requirements of the Part 178.3297 'Colorants for polymers'.

   Expert advice would be required to ensure the correct choice was made, but it is possible to use a 'read across' approach, if it is appropriate in terms of compatibility, long-term stability, type of food to be contacted and conditions of contact, it may be possible to use a pigment in the PET that is present in other sections of CFR Title 21, such as:
   - Colour additives listed for direct use in food itself
     - Parts 73, 74, 81 and 82
   - Paper and Board: Part 176.170
   - Resinous and polymeric coatings: Part 175.300
   - Generally regarded as safe (GRAS) substances (e.g., iron oxide): Parts 184 and 186

3. *Other additives in the PET product*: For it to perform satisfactorily during processing and in service, additives such as antioxidants and ultraviolet (UV) stabilisers may have to be added to the PET product. Any additive present must be compliant with the regulations by being present within an appropriate section of the CFR.

   For example, any stabilisers must be listed in Part 178.2010 – Antioxidants and stabilisers for polymers, or another appropriate section (e.g., GRAS – Parts 184 and 186).

4. *Migration testing of the PET product*: In some parts of CFR Title 21 (e.g., Part 177.1520 – Olefin Polymers) there are no requirements for migration testing, but some suppliers and manufacturers carry out migration testing under applicable end- use conditions using the testing protocol outlined in other Parts of Title 21 (e.g., Sections 176.170 or 175.300).

   However, migration tests are included in the PET polymers section, Part 177.1630, and a PET product must meet the requirements of the tests applicable for its intended end-use (e.g., particular food type and contact conditions). An example illustrating the testing required is provided below.
   - *Intended end-use of the PET product*: Intended to transport, package or hold food, excluding alcoholic beverages, at temperatures ≤121.1 °C.
   - *Migration testing that must be carried out*:
     - The food contact surface, if exposed to distilled water at 121.1 °C for 2 h, should not yield chloroform soluble extractives >0.5 mg/in^2; and
     - The food contact surface, if exposed to n-heptane at 65.5 °C for 2 h, should not yield chloroform soluble extractives >0.5 mg/in^2.

With respect to using recycled plastics for manufacturing food-grade products, there is usually no mention of their use to manufacture food contact materials in CFR Title 21 and this is true for the PET section, Part 177.1630.

To use rPET and other recycled plastics to produce food contact products it may be necessary to get a 'letter of no objection' from the FDA. This is achieved by submitting a dossier to the FDA that demonstrates that the product is food-safe. A list of these '*Submissions on Post-Consumer Recycled Plastics for Food Contact Articles*' is available on the FDA website [19]. Each submission is given a unique number (a recycle number) and the date of the 'no objection letter' is provided along with the name of the company that had made the submission, the polymer that was the subject of the submission, the recycling process, and any restrictions that apply to the end-use of the recycled plastic. At the time of writing, 187 submissions have received a 'no objection letter' from the FDA, with most being for processes that produce food-grade rPET.

Some of these letters have been reported as being issued as far back as the early-1990s. Letters have been issued for a least three generic types of rPET product/process: chemical depolymerisation of rPET into monomers for vPET production; food contact products produced from 100% rPET; multi-layer materials in which an rPET layer is combined with a vPET food contact layer. Information of this type can be sourced in a report entitled '*Best Practices and Industry Standards in PET Plastics Recycling*' [20] available on the US National Association for PET Container Resources (NAPCOR) website [21] (Section 3.2). The contents of this guidance document are discussed in Section 4.3.4.

4.3.4 Guidance documents for recycling food-grade plastics in the US

There are FDA guidelines for the formal approval of recycled plastics in direct food contact applications. These guidelines are based on the 'threshold of regulation' concept and involve the testing of recycling processes in a similar way to those described in EU Recycling Regulation (EC) 282/2008. The challenge tests used in the FDA guidelines involve 'surrogate' substances that mimic the potential abuse of plastic products in service and the mixtures used are designed to test the ability of the recycling process to remove them down to a 'safe level'. The chemical substances used are similar to those employed in the EU tests, and the FDA have set a general permissible migration level of 0.5 ppb per surrogate substance. This is the level that is defined in their concept of threshold of regulation, but it varies according to the 'consumption factor' of a specific plastic. The consumption factor takes into account the diffusion behaviour of a plastic as well as its market share, and its use in the case of rPET increases the migration limit to 10 ppb per surrogate substance.

This guidance is available to industry from the FDA in the form of a document: '*Use of Recycled Plastics in Food Packaging: Chemistry Considerations*'. The present version of this document, published in 2006, supersedes the document published in 1992, which was entitled '*Points to Consider for the Use of Recycled Plastics in Food Packaging*'. As with the EU regulation, this FDA guidance document addresses the possibility that chemical contaminants in plastic materials intended for recycling may remain in the recycled material and could migrate into the food that the material comes into contact with. This is one of the main considerations for the safe use of recycled plastics for food contact applications, but does not address other important aspects, such as microbial contamination and important bulk properties (e.g., tensile strength). The FDA are keen to point out that their guidance documents (including this one) do not establish legally enforceable responsibilities but instead describe the FDA's current thinking on a topic, and should be viewed only as recommendations, unless specific regulatory or statutory requirements are cited. This document is a useful source of information on the subject and covers many areas of interest:

- Definition of different types of recycling processes:
  - *Primary recycling*: Re-use of in-house food-grade manufacturing scrap.
  - *Secondary recycling*: Physical recycling that involves processes such as grinding and decontamination (e.g., by washing).
  - *Tertiary recycling*: Chemical recycling by depolymerisation to starting materials (i.e., monomers) and then re-polymerisation to a new polymer.
- Exposure to chemical contaminants from recycled plastics.
- Surrogate contamination testing of recycling processes.
- Presence of non-food approved plastic in the waste stream.
- Use of effective barriers (i.e., functional barriers) in multiple-layer products to enable non-food approved recycled plastic to be used on the non-food contact side.

To demonstrate that a secondary or tertiary recycling process (see above) can be used to remove contaminants from post-consumer plastic, the FDA recommends that consumer misuse should be simulated by exposing virgin polymer (in container form or as flake) to selected surrogate compounds and then running this material through the process and testing the resulting recycled polymer. This subject is covered in the section on testing of surrogate contamination (Section V). The document provides guidance on the surrogate compounds that should be used to simulate consumer misuse, and states that the FDA believe that one surrogate from each of the categories shown in Table 4.3 is sufficient for this test.

Once the containers are filled, or a mass of flakes mixed thoroughly with the cocktail, they should be stored for 2 weeks at 40 °C with periodic agitation. Then, the cocktail is removed from the containers (or flake) they rinsed to remove residual cocktail from the surface. An analysis is then done to determine the concentration of each surrogate compound in the polymer ('challenged polymer') and this is run through the recycling process, and packaging materials made from this recycled polymer are analysed to determine the concentration of residual surrogates. This approach, in common with the EU method (Section 4.2.6), represents a worst-case scenario because it assumes that all the post-consumer plastic entering the recycling stream is contaminated.

With regard to employment of an effective barrier for use of recycled plastic from a secondary or tertiary process for the manufacture of food contact materials (Section VII), the FDA has determined that vPET is an effective barrier to contaminants that could migrate from a layer of recycled plastic under the following conditions:

- At a thickness of ≥25 µm at room temperature and below, and
- At a thickness of ≥50 µm at higher temperatures, including use as a dual-ovenable container for cooking food at 150 °C for 30 min*.

  *Providing that only food containers are used in the feedstock to manufacture the recycled layer.

In the US, there are specific guidance documents written by consultants for government departments. An example of such a document specific to PET is a best-practice and industry standards guide written by Bronx 2000 Associates in 2002 [20] for the Washington State Department of Community, Trade and Economic Development. The full title of this document, which covered the recycling of all types of PET (food and non-food-grades) is '*Best Practices and Industry Standards in PET Plastic Recycling*' (Section 4.3.3). As well as providing an overview of PET recycling and the issues around permits and regulations at PET-recycling facilities, it covers two principal stages in the recycling process:

- Best practices in PET collection
- Best practices in PET intermediate processing

The document, therefore, covers the PET-recycling process up to where the post-consumer PET has been collected, sorted, baled and ground into unwashed flake.

References

1. http://ec.europa.eu/environment/ipp/integratedpp.htm
2. *Global Legislation for Food Packaging Materials*, Eds., R. Rijk and R. Veraart, Wiley-VCH Verlag GmbH & Co., Weinheim, Germany, 2010.
3. *Global Legislation for Food Contact Materials*, Ed., J.S. Baughan, Volume 278 of Woodhead Publishing Series in Food Science, Elsevier Science and Technology, Amsterdam, The Netherlands, 2015.

4. R.S. Baxi in *Recycling our Future: A Global Strategy*, Whittles Publishing, Dunbeath, UK, 2014.
5. *Resource Recovery to Approach Zero Municipal Waste (Green Chemistry and Chemical Engineering)*, Eds., M.J. Taherzadeh and T. Richards, CRC Press, Boca Raton, FL, USA, 2015.
6. http://www.efsa.europa.eu
7. http://www.efsa.europa.eu/en/press/news/120802.htm
8. http://www.ec.europa.eu
9. *The EFSA Journal*, 2008, **717**, 1.
10. T.H. Begley, T.P. McNeal, J.E. Biles and K.E. Paquette, *Food Additives and Contaminants*, 2002, **19**, 135.
11. R. Franz, F. Bayer and F. Welle in *Guidance and Criteria for Safe Recycling of Post Consumer Polyethylene Terephthalate into New Food Packaging Applications*, Report EUR 21155, Office for Official Publications of the European Communities, Luxemburg, 2004.
12. R. Franz, F. Bayer and F. Welle in *Guidance and Criteria for Safe Recycling of Post Consumer PET into New Food Packaging*, Report of the EU Project FAIR-CT98-4318 'Recyclability', Section I: PET Recyclability, February 2003.
13. *The EFSA Journal*, 2011, **9**, 7, 2184.
14. *Recycling of Plastics for Food Contact Use: Guidelines*, International Life Sciences Institute (ILSI) Europe Report Series, ILSI Press, Washington, DC, USA, May 1998.
15. http://waste-management-world.com
16. *Packaging and Environmental Legislation in the United States: An Overview*, Keller and Heckman LLP, Washington, DC, USA, 2015. http://packaginglaw.com
17. http://www.epa.gov/recycle
18. http://www.accessdata.fda.gov/scripts/cdrh/cfdocs/cfcfr/ *cfrsearch.cfm*
19. http://www.accessdata.fda.gov
20. D.J. Hurd in *Best Practices and Industry Standards in PET Plastic Recycling*, Bronx 2000 Associates Inc., Bronx, NY, USA, 1997.
21. http://www.napcor.com

5 Separation and sorting technologies

5.1 Introduction

To achieve the highest possible quality, it is crucially important to separate different plastic types from one another because the difference in their melting/softening points and other fundamental properties (e.g., chemical and thermal resistance) mean that they are incompatible. The importance of ensuring that the sorting of mixed plastic waste is carried out effectively has been demonstrated by studies that have analysed the quality of the recyclate produced by municipal solid waste (MSW) facilities. Such a study was carried out in The Netherlands [1] and the results showed that improvements to the sorting systems improved the quality of the plastic recyclate and so improved consumer confidence and increased market impact.

The influence that separation problems have on the amount of plastic that must go to landfill has been highlighted by Waste and Resources Action Programme (WRAP) in the UK. In 2011, WRAP reported [2] that despite >300,000 tonnes of plastics packaging being collected and recycled in the UK each year, more than 1 million tonnes ended up in landfill because of several difficulties: collecting and recycling plastic films, detecting and sorting black plastics, and the lack of high-value markets for non-bottle plastic waste. However, WRAP was optimistic that the situation could improve soon due to the research that it was sponsoring, for example, the trials that it had sponsored with non-carbon black pigments that demonstrated that black products could be sorted using standard near-infrared spectroscopy (NIR) sorting equipment (Section 5.3.3).

It is also vital to separate the overall plastic 'stream' from non-plastic materials such as:
- Metal (e.g., from screw caps)
- Paper and glue (e.g., from labels)
- Coatings and inks (e.g., from printing)
- Dirt collected during storage and transportation
- Residual product (e.g., food)

These types of contaminant were highlighted by a report published by WRAP in the UK [3] and, to achieve a high-level of purification, post-consumer polyethylene terephthalate (PET) must go through many processes. Several photographs were published in this report illustrating the separation tasks to be tackled to obtain a high-quality recycled polyethylene terephthalate (rPET) product (Figures 5.1–5.4). These figures show typical PET tray packaging showing labels, a mixture of plastic flakes from post-consumer plastic, granulated flakes with labels before several air-classification and washing steps, and coloured particles in PET flake ground to <2 mm in size.

https://doi.org/10.1515/9783110640304-005

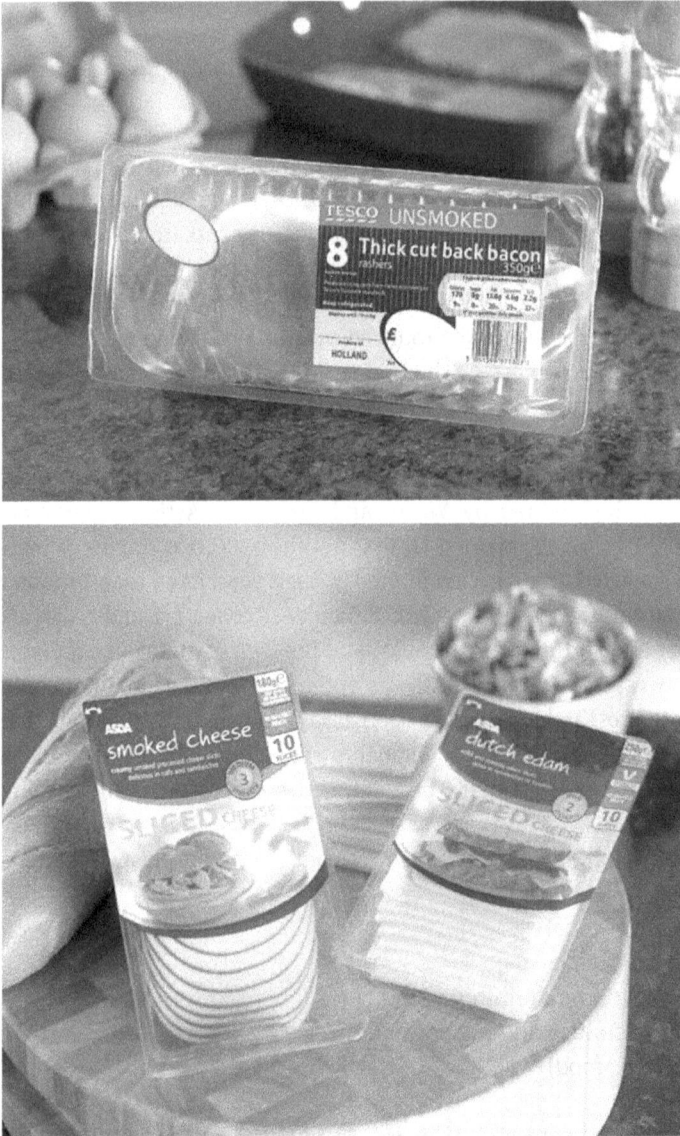

Figure 5.1: Typical PET tray packaging in the UK showing labels, and lidding films. Reproduced with permission from the Waste and Resources Action Programme (WRAP), Banbury, UK. ©WRAP.

Recycling facilities can operate in different ways according to whether they recycle only PET or accept mixed plastic waste, but several of these stages are common to both, and a generic sequence of preparation, separation and decontamination steps can described. The sequence shown below is for PET products (e.g., bottles) that

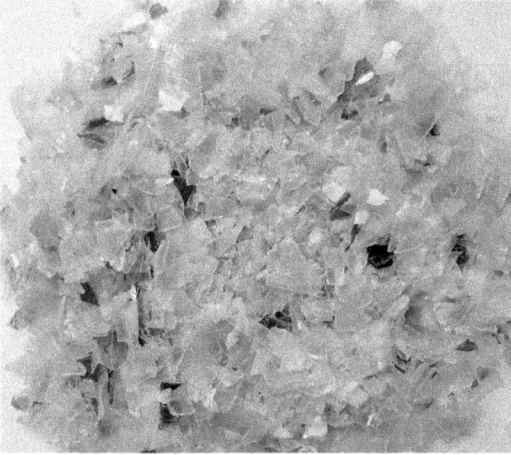

Figure 5.2: Mixture of plastic flakes from post-consumer plastic. Reproduced with permission from the Waste and Resources Action Programme (WRAP), Banbury, UK. ©WRAP.

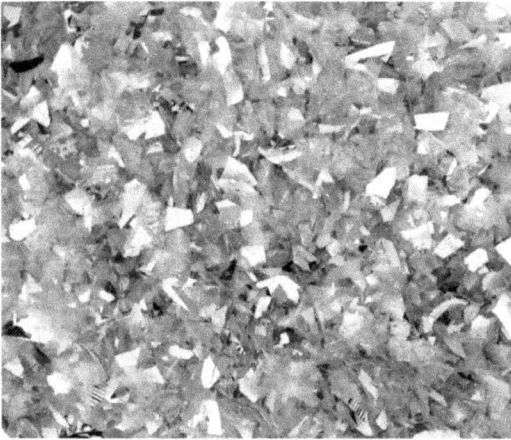

Figure 5.3: Granulated flakes with labels before several air- classification and washing steps. Reproduced with permission from the Waste and Resources Action Programme (WRAP), Banbury, UK. ©WRAP.

have been separated from other types of plastics and non-plastic items (e.g., fabric and metal cans) by use of NIR detection and other separation techniques. The separated PET products will be subjected to several processes:

1. Initially the fraction will be placed into a granulator to produce flakes ≈25 mm in diameter.
2. Then, the flakes can pass through a caustic (i.e., alkali) hot-wash process to remove surface dirt, paper-label fragments and label adhesive.

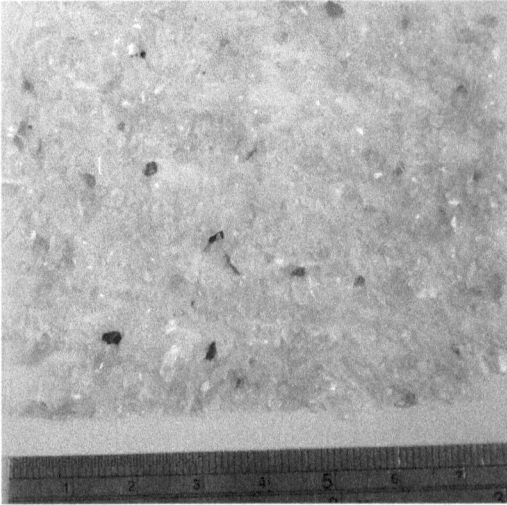

Figure 5.4: Coloured particles in PET flake ground to <2 mm in diameter. Reproduced with permission from the Waste and Resources Action Programme (WRAP), Banbury, UK. ©WRAP.

3. To ensure the highest possible level of decontamination, the PET flakes can go through set of decontamination processes which typically comprise a metal detector, a colour sorter, and a series of polymer sorters (e.g., NIR and Raman types). Then, some recyclers employ a chemical-cleaning process to remove contaminants absorbed by the PET in service.

PET flake that fully completes such a recycling sequence is then usually labelled as 'high-quality washed flake' (HQWF) and can be sold as a commercial product (e.g., to make PET strapping products).

To convert the HQWF into food-grade PET it must be purified to a higher degree using one of the many commercial processes (Chapter 6) shown to decontaminate the PET to the extent that the material will meet the requirements of the US Food and Drug Administration (FDA), European Food Safety Authority (EFSA) and the European Union (EU) Plastics Recycling Regulation, (EC) 282/2008. The recycler must demonstrate this capability by carrying out a 'challenge test' on their recycling process and the PET flake produced by it, and by running the process within an acceptable quality control [i.e., good manufacturing practice (GMP)] regimen. The EFSA have published guidelines on how to carry out such challenge tests, and a quality system such as International Organization for Standardization (ISO) standard, ISO 9001 is usually sufficient to meet GMP requirements (Chapters 4, 6 and 7).

The fact that the separation and sorting of plastic waste is a crucial stage in the recycling process has also been highlighted in studies such as the one published by Luijsterburg and Goossens on the recycling of packaging waste [1]. They showed that the collection method for plastic packaging waste hardly influences the final quality of the recyclate, but that the sorting and processing steps have an influence. They also demonstrated that, though the mechanical properties of the recyclate are different from those of virgin polymers, changes to the sorting and reprocessing steps can improve its quality.

Industry is constantly seeking quicker and more efficient ways to deal with plastic waste products. Kluenker [4] wrote a report describing trends in the conversion of plastic waste into useable materials. He discussed the sorting of plastic waste, reduction of this waste to a uniform particle size and shape, and removal of impurities. Other reviews have assessed and summarised the progress being made in the sorting and separation of post-consumer plastic. One example of such a review was presented in a paper by Schuh [5] at a recycling conference. It described the work carried out by ASCON to assess the state of the art of the automatic sorting of waste using sensor-based NIR separation and eddy current separation. It also described the progress in various parts of the world towards minimising the amount of waste going to landfill and maximising recycling. The effectiveness of the various sorting techniques was commented on, as was the economic return for plastics and metals from the investment made in them. The productivity and efficiency for the production of homogeneous, marketable secondary raw materials for international markets were also assessed. In another review, Kosior and Dvorak [6] addressed the performance of advanced separation techniques applied to the recycling of mixed plastic waste streams generated from post-consumer collections after the main bottle fraction had been removed. The composition of the plastics, process costs and economic value of the finished products were also evaluated to ascertain if these techniques provided a viable method of recycling mixed plastic waste streams.

In common with many of the major stages in the recycling process, considerable research is ongoing on sorting and separation. Much of this research is targeted on tackling:
- Identification and separation of black and highly coloured plastic items.
- Differentiation between food-grade and non-food-grade products of the same generic polymer type [e.g., polypropylene (PP)].
- Removal of metal-catalysed oxobiodegradable plastics from waste streams.
- Removal of biodegradable plastics [e.g., polylactic acid (PLA)] from waste streams.

These areas are addressed in Section 5.3.

5.2 Sorting and separation technologies for plastics

5.2.1 Spectroscopic separation technologies

As mentioned in Section 5.1, it is possible to carry out separation of PET from other plastics while they are in their original product form (i.e., bottle or tray) or, after it has been through a grinding process and been converted into flake, by use of a conveying system (e.g., a conveyor belt) and a detector, such as a NIR detector and/or a Raman spectroscopy detector. These types of detection system, particularly the NIR detector, are regarded as industry standards, but it is also possible to use other types of detection system for this operation. A summary of the different types that are available commercially or which have been evaluated by research projects, such as the EU FP7-funded project SuperCleanQ, [7] and that conducted by WRAP [8], are shown in Table 5.1.

Table 5.1: Different spectroscopic detection technologies.

Technology	Comments
NIR	One of the industry standards and its commercial use is widespread. Research has shown that it is also capable of identifying multi-layer materials and some additives
Raman-IR spectroscopy	One of the industry standards and its commercial use is widespread. Research has shown that it is also capable of identifying multi-layered materials and some additives
Reflectance MIR spectroscopy	Systems are used commercially for high-speed sorting of flake but not used for complete articles
Laser Raman fluorescence spectroscopy	Similar to MIR with systems used commercially for high- speed sorting of flake but not for complete articles
X-ray fluorescence spectroscopy	Works for polymers containing elements of high atomic number (e.g., chlorine in PVC) but does not work well for elements of low atomic number (e.g. H, C and O) – the only elements that most common polymers contain
Ultraviolet– visible spectroscopy	Research has shown its potential to be used for some minor specialist applications (e.g., identification of certain additives), but it can be affected by the colour of the component and dust. Spectral data not as specific as those for NIR or IR spectroscopy
LIBS/LIPS	Has good capabilities and has potential to increase in importance commercially
Photoacoustic spectroscopy	Tested for recycling automotive and electronic waste but not known if robustness and response time is adequate for recycling plastic packaging

IR: Infrared
LIBS: Laser-induced breakdown spectroscopy
LIPS: Laser-induced plasma spectroscopy
MIR: Mid-infrared spectroscopy
PVC: Polyvinyl chloride
Reproduced with permission from the SuperCleanQ EU-funded FP7 Research Project, http://www.supercleanq.eu. ©SuperCleanQ [7] and *Development of NIR Detectable Black Plastic Packaging*, Final Report, Waste and Resources Action Programme (WRAP), Banbury, UK, September 2011. ©2011, WRAP [8]

Some of the techniques described in Table 5.1 have been investigated to evaluate their potential for the identification and separation of black plastic items from the recycling stream, and this application is discussed in Section 5.3.

An article by Shelley [9] described how the development of NIR laser diodes and detectors has enabled the rapid recognition and reliable separation of six plastics simultaneously. Specifically, a five-diode laser sensing system developed by the Japanese company IDEC was featured. It was claimed that the wavelength produced by the IR laser diode could be changed by varying its composition and fabrication. The review stated that, by using a combination of these laser diodes operating at different wavelengths it was possible to distinguish between high-density polyethylene (HDPE), low-density PE, polycarbonate, PET, PP, polystyrene (PS) and polyvinyl chloride (PVC).

In an article in *Plastics and Rubber Asia* [10] it was reported that S+S Inspection Asia offers several sorting systems, including a NIR sensor for the sorting and removal of bioplastic materials, such as labels made from polylactic acid (PLA) or polyethylene terephthalate glycol, from PET bottle streams.

5.2.2 Other types of sorting technologies

It is also possible to granulate mixed plastic waste and then use other techniques to separate one plastic from another, using intrinsic properties such as density and chemical structure. These non- spectroscopic techniques include:
- Density-based techniques:
  - Sink–Float methods
  - Windshifting techniques
  - Hydrocyclone methods
- Electrostatic methods

A brief summary of each of these techniques is provided in Sections 5.2.2.1 and 5.2.2.2.

5.2.2.1 Density-based techniques

5.2.2.1.1 Sink–float methods
Sink–Float methods are the most commonly used methods to separate plastics mixtures for recycling and allow separation of a mixed composition into two factions according to the differences in density of the component polymers so that the variations in the size and shape of the plastic waste have very little influence on the separation process. The sink–float process is usually carried out in a tank that contains drive rollers to agitate the lighter, floating material located at the top, with an

outlet for the denser component placed at the bottom of the structure. This outlet can be connected to a conveyor screw to transport the separated fraction on to the next stage in the process, which may involve another separation technology to refine the separation process further. For example, recyclers often complement sink–float tanks with additional systems such as friction washers. These can be used to remove contaminants (e.g., cellulose) and consist of a densely perforated cylinder with 2-mm holes and a propeller-driven internal conveyor–friction system that fits closely within this cylinder. The plastic particles are mixed with water as they come from the sink–float system and are moved forward by the propeller and pressed against the walls of the perforated cylinder. Dirt of diameter <2 mm, together with paper and cellulosics that break into fibres and water, go through the holes in the cylinder and are removed through the outlet. The propeller drives the wet, cleaned plastic particles through the other outlet at the end of the cylinder. The water used in the process is recovered and sent back to the mixing tank after being cleaned by filtering.

Usually, the separation fluid is water to which some surfactant is added to avoid flotation of denser particles due to surface tension. Such lines are very useful for separating polyolefins because they have a lower density than most plastics, which enables them to float in water. If these systems are used to separate PET, it will be collected at the bottom of the tank due to its relatively high density ($\approx$1.38 g/ m^2). The steps before and after the sink–float tank vary depending on the composition of the starting mixture and the final objective of the process.

5.2.2.1.2 Elutriators and other variations of sink–float methods

Elutriators and other variations of sink–float methods are also based on water and consist of a vertical tank in which the sink–float technique is combined with an input of water of a variable speed from the lowest part of the system. Therefore, with the elutriators there are two opposite forces: one that tends to sink particles of higher-density than the media fluid and one than tends to float them *via* the push towards the surface by the water jet. By the use of elutriators, a separation in three streams can be undertaken: a stream of floating particles on the surface, another steam of heavy particles at the bottom of the tank, and a third stream in the middle of the system consisting of particles of medium density. However, with elutriator methods, the size and shape of the particles can play important parts in the process. Therefore, the separation obtained can be less accurate than that for traditional sink–float processes.

Other variations of sink–float processes include using a separation fluid of a density greater than water (e.g., solutions of salt water). In this way, it is possible to float plastics (e.g., PET) that are denser. Another variant is froth flotation, and Carvalho and co-workers [11] have reported on optimisation of a 'froth flotation' procedure for separation of PET from PVC and PS. They carried out the study using representative samples of post-consumer waste and, before any attempt was made at separation, the PET in the mixture was wetted selectively by alkaline treatment followed by surfactant adsorption. Following the separation, an enriched product with 98.9% PET and only 0.6% PVC was obtained in the non-floated product. With regard to the final recovered fractions, the result for PET from the non-floated product was 97% whereas, from the floated product, the recovery of PVC was 97%, with 91% for the PS. The research team also developed an analytical method based on selective dissolution for quantification of the composition of the waste plastic mixture.

5.2.2.1.3 Windshifting techniques
Windshifting techniques also rely on components within the mixed plastic waste having different densities. Several designs are available commercially, including Vertical, Diagonal and Zig-Zag Shifters. These designations refer to the overall configuration of the various components in the design, but typically they all consist of a re- circulation fan, a separation unit and a combi-separator. Windshifting techniques offer benefit over sink–float methods by enabling a higher throughput of waste (e.g., capacities ≤100 tonnes/h) to be achieved. Some of the other benefits of windshifters include:
– High-separation efficiencies (≤99%)
– Low-maintenance and very few wearable parts ensures less downtime
– Low-operational costs
– Low-dust emission

5.2.2.1.4 Hydrocyclone methods
Hydrocyclone methods have gained market share from sink–float methods because they can achieve separations which are as accurate, but have the advantage of being faster, more compact, and being able to separate plastics that are slightly denser than water without having to vary the separation fluid. The drawback with these systems is that they are more expensive and, therefore, usually applied only to the separation of large amounts of plastic waste. Furthermore, the size and shape of the plastic particles can influence the quality of the separation that is achieved.

The hydrocyclone method is based on introduction into the hydrocyclone of a transport fluid (usually water) that carries the plastic mixture to be separated. The plastic mixture comprises particles that are usually within the size range 5–10 mm. This mixture moves down in a spiral trajectory close to the walls of the hydrocyclone until reaching the bottom conical piece (the apex) that has a purpose-designed geometry and evacuation diameter. On arrival at the apex, part of the fluid is evacuated, carrying the heavy material, and producing simultaneously an internal ascending flow that exits through the opening at the top of the hydrocyclone, carrying with it the lighter fractions of the mixture. A high level of success has been achieved by use of hydrocyclone systems to separate post-consumer HDPE bottle and PET bottle mixtures. In this case, the HDPE would exit the top of the hydrocyclone and the PET the bottom. Friction washers can also be used within hydrocyclone-based separation systems for removal of paper fibres and other contaminants.

The design of a hydrocyclone process for the separation of plastics is complex and the variables that must be considered will define a 'cut point' (i.e., which conditions the plastic will descend or ascend in the primary and secondary streams). Very precise adjustment of the conditions is required if the densities of the plastics to be separated are close together. Fine adjustment of these conditions is also used to obtain an optimal compromise between the maximum amount of plastic recovered and the maximum level of purity attained. Despite the high selectivity of these systems, they are not usually capable of separating PET from plastics that have very similar densities (e.g., PVC) and, in these cases, other separation systems (e.g., spectroscopic) must be used.

5.2.2.2 Electrostatic separation

Electrostatic separation is based on the fact that insulating materials, such as plastics, can be charged readily through an electric discharge (corona charge) or through friction at their surface (triboelectric charge). The more conductive a plastic, the faster it will dissipate the charge introduced, returning the plastic to its neutral state. Plastics with low conductivity will maintain their charge for a relatively long period of time and so will be attracted or repulsed by a charged electrode.

The differences in the dielectric constants of plastics are not high enough to enable use of a separation system based on disappearance of the induced electrostatic charge. However, mechanisms based on triboelectric charge are feasible because the magnitude of the charge will depend on the dielectric characteristics of the plastic and hence on its chemical composition. Different plastics can, therefore, be separated into different streams by the use of high-voltage electrodes.

One of the drawbacks of these systems is that the quality of the separation is influenced by the temperature and water content of the plastic and the environment. These parameters make necessary placement of the system in an area with control of temperature and humidity and pre-conditioning of the plastic mixture to achieve highest-quality separations. Another factor that can affect the quality of the separation adversely that is obtained is surface contamination on the plastic particles. These systems have been considered for the separation of black plastic items, but these drawbacks have been thought to work against them (Section 5.3.1 and Table 5.2).

5.2.3 Removal of non-plastics from the recycling stream

Contamination, such as metal fragments, can be present in a plastics recycling stream for many reasons (e.g., small fragments from the collars associated with

Table 5.2: Different detection technologies for the sorting of black plastics.

Technology	Capability
Reflectance MIR and laser Raman fluorescence spectroscopy	These two technologies are regarded as having potential for the future in this area but at present there are some limitations with the equipment available
X-ray fluorescence spectroscopy	Limitation regarding ability to detect C, H and O as described in Table 5.1
LIBS/LIPS	Promising development for use in the future but not used commercially currently
Photoacoustic spectroscopy	Yet to be evaluated for recycling of plastic packaging – see Table 5.1
Electrostatic separation	Has potential to provide adequate separation but surface contamination can interfere with it
Tack separation	Will not supply sufficient separation for complex mixtures and so not applicable for this application
Alternative black pigments to carbon black	Black colourants with lower NIR absorption* and those that reflect NIR** may work with reflectance NIR detectors but not in transmission. Both types provide black colouration without absorbing NIR strongly. Food contact compliance must be considered and additional cost must be justified and accepted by industry
Addition of markers, such as fluorescent additives, taggants, and RFID tags	Fluorescent markers could provide a cost- effective method of identification, but the addition rates to work effectively in black plastics must be established. Some problems are associated with other types of markers

* e.g., BASF Lumogen and trichromic mixtures or colourants or dyes
** e.g., BASF Sicopal K0095
RFID: Radio-frequency identification tags
Reproduced with permission from *Development of NIR Detectable Black Plastic Packaging*, Final Report, Waste and Resources Action Programme (WRAP), Banbury, UK, September 2011.©2011, WRAP [8]

metal screw tops). Metal contamination must be removed from plastics recycling lines to stop it damaging the down-steam equipment (e.g., pellet extruders) and spoiling the recycled products. Metal can be removed by magnetic systems and several systems are available commercially. For example, when a mixed plastic bottle sorting facility of 140,000 tonnes/year was set up at Hemswell, Lincolnshire, UK, by a Eco Plastics–Coca Cola Enterprises joint venture, a problem was encountered with the detergent bottle line because many were fitted with metal spring-loaded trigger mechanisms. To address this problem Eclipse Magnetic provided a 300-mm

Auto Rota Shuttle magnetic separation system that could extract all ferrous contamination, including fine particles, from granulated PET moving at 5 tonnes/h [12].

5.3 Separation of black and highly coloured plastic

5.3.1 Introduction

The limitations associated with standard industry detectors (e.g., NIR types) for the separation of black and highly coloured plastic from post-consumer plastic streams is a serious problem for the recycling industry (Figure 5.5). This problem occurs because the standard pigment used in the polymer industry to colour items black is carbon black, which absorbs most of the NIR radiation from the NIR

Figure 5.5: Examples of black plastic packaging found in mixed plastic waste residue streams. Reproduced with permission from the Waste and Resources Action Programme (WRAP), Banbury, UK. ©WRAP.

detectors, so there is insufficient reflected back to the detector to enable it to identify the plastic by reference to the spectral information in its database. Carbon
black is an attractive pigment for several reasons: it is relatively cheap and can
create a good black at small levels of incorporation (e.g., <1% w/w).

The ability to separate and recover black plastic would boost the economic efficiency of recycling operations because of improved yields, lower levels of residual
waste to landfill and the opportunity for greater revenues from the sale of recycled
resins produced from the recovered black plastics. The multiple-faceted sorting systems (Section 5.1) used in the plastics recycling industry are essentially the same
whatever polymer type is being targeted. Therefore, the problems that exist regarding
black-coloured items due to the use of these systems are also common to the whole
industry. Hence, any improvements that can be made in this area could be exploited
throughout it, and not just be confined to the sorting of one particular (e.g., PET)
plastic waste stream.

In the case of plastic packaging, the key polymers used in this sector are PP, crystalline polyethylene terephthalate (CPET) and amorphous polyethylene terephthalate
(APET), the three making up ≈80% of the market, with HDPE, PS and PVC making up
the remainder. The recycling and reprocessing issues associated with each polymer
are quite different once they are separated into their specific coloured plastic fraction.
The markets for recycled black PP and black PET are quite different, but both plastics
are valuable and can be used in many applications. For PET, there are potential end
markets for mixtures of APET and CPET, as well as for separate APET and CPET
streams. Once the fundamental problem of the presence of the black pigment has
been solved, and the polymer response is detectable, it should be possible to differentiate between the APET and CPET forms of PET (Figures 5.6 and 5.7) [13].

Some of the potential markets for APET and CPET are summarised below [13]:
- *APET*: The end markets for coloured APET are typically fibre and strapping. The
  fibre market uses APET for textiles (which could be coloured dark shades) or as
  roofing membranes (where it is covered with bitumen). In both cases, the presence
  of black PET would not be a problem. It could also assist savings by reducing the
  amount of black masterbatch that needs to be added if a dark or black colour is
  required and when the black rPET comes around to being recycled itself.
  Once it is possible to separate black APET items, it might be possible to develop
  a specific rPET-designated black strapping product that may have an economic
  advantage for strapping companies.
- *CPET*: Trays tend to be made from CPET and so this would enable black PET
  trays to be recycled. The presence of detectable black APET would not be a
  problem in this fraction because it would not interfere with its re-use, and
  would increase its volume.
  This would enable the loop to be closed on CPET trays back to food-grade for
  the very first time, and allow this significant volume of material to be recovered
  from the waste stream.

Figure 5.6: CPET trays sorted by NIR after a single pass. Reproduced with permission from the Waste and Resources Action Programme (WRAP), Banbury, UK. ©WRAP.

Figure 5.7: Non-CPET products separated from black CPET trays after a single pass. Reproduced with permission from the Waste and Resources Action Programme (WRAP), Banbury, UK. ©WRAP.

At lower recovery rates, the black CPET could be directed into the black APET waste stream for use in fibres or strapping, where it would not cause a problem, whereas it would in the bottle or sheet market due to its colour.

Taking the situation in one European country, the UK, the total amount of black plastic packaging in the UK market has been estimated by WRAP [8] as being between 25,000 and 60,000 tonnes/ annum. This fraction of black plastic (1.5–3.5%) is present in the 1.7 million tonnes of household packaging waste, the main component of plastic waste collected for recycling.

With regard to the EU economic area, data produced by Eurostat in October 2013 [14] showed that the total amount of plastic packaging waste in the EU was ≈15 million tonnes. Assuming the same proportion of black waste exists in this waste as in the UK market (i.e., 1.5–3.5%), a total figure for the EU for black plastic waste of 225,000–525,000 tonnes is provided.

A research programme by WRAP [8] showed that even if PET products (e.g., food trays) were manufactured using black UN pigment (a carbon black-type pigment) at levels as low as 0.1% w/w, a high-intensity NIR detector (43% higher than the standard intensity) failed to identify the polymer type. The industry is trying to overcome these limitations in the detection and sorting systems so that this 'black fraction' can be accessed and recycled. It has made some progress, and this is summarised in Table 5.2 but no complete solution has yet been found.

The technologies listed in Table 5.2 fall into the following categories:
1. Use of alternative spectroscopic techniques
2. Use of alternative black colourants to carbon black that can be detected by NIR
3. Use of physical techniques that do not involve spectroscopy-based methods
4. Modification of the plastic by the use of marker compounds

Due to factors that affect the last two categories (e.g., technical and commercial restraints) the first two categories are regarded as having the best possible chance of finding a solution to the black- plastic problem. Of these two possibilities, the approach using an alternative black colourant tends to have more supporters because this could be effective without the need to change the current NIR detection systems used at existing plastics recycling sites. Some of the developments taking place in all four areas are discussed in Sections 5.3.2–5.3.5.

5.3.2 Developments in spectroscopic systems

The inherent difficulty with NIR-type sorting systems was highlighted in a WRAP research project [15] aimed at the evaluation of different sorting systems for the separation of different types of plastic components in mixed waste electrical and electronic equipment waste streams. One of the NIR detectors trialled by WRAP could differentiate between different polymer types in light-coloured products, but could not identify 50% of the waste plastic because it was too dark. The other NIR detector used in the trial also had difficulties in identifying dark-coloured plastics.

New detection systems are appearing in this market, but companies are also working to improve the performance that can be achieved using existing NIR technology. For example, the performance of new NIR-type sorting detectors developed by Pellenc ST was evaluated using mixtures of plastic electrical waste [16]. The conclusions were positive and the new devices were regarded as offering viable commercial alternatives to systems presently on the market.

5.3.2.1 Mid-infrared spectroscopy

Mid-infrared spectroscopy (MIR) is used extensively for the identification of polymer type in a laboratory, but it suffers from two main deficiencies for automated sorting. The time of detection is ≥1 s, which is slower than with NIR and the computational requirements to match the product spectrum to reference spectra are also more complex due to the many peaks in the MIR 'fingerprint region'. An advantage is that absorption by carbon black is less strong in the MIR region than for NIR or visible regions [17]. MIR spectroscopy can identify black plastics for this reason, though there are difficulties for industrial applications due to the need for close proximity of the sample and detector [18], and sensitivity to the surface characteristics of the articles to be sorted [19].

MIR spectroscopy has found use in identification of materials in automotive parts, but does not appear to be viable for use in high- speed automated sorting of post-consumer plastic packaging without significant equipment development. At the moment, although it has attractive technical attributes, it is not regarded as a serious competitor technology in this area.

5.3.2.2 Raman spectroscopy

Raman spectroscopy is a potential alternative to NIR spectroscopy that has received considerable attention in plastics sorting. It is an emission technique that does not rely on the measurement of absorbed or reflected radiation. A laser operating in the NIR or visible region is used to excite the material, producing a characteristic emission spectrum in the IR region through the Raman scattering effect.

The potential of Raman spectroscopy for plastics identification in recycling operations has been known for a long time [20] but, as with NIR spectroscopy, black pigmented samples cannot be measured, due to high-surface absorption of the incident laser pulse and extra fluorescence from the carbon black. Recent advances have been made to solve this problem, but this equipment is targeted for automotive and electronic waste rather than sorting of post-consumer packaging.

Sony claim that the technological advances that they have made make Raman spectroscopy well suited to plastics sorting, including black plastics [21]. Also, the Unisensor POWERSORT 200 flake- sorting system uses laser Raman fluorescence spectroscopy and is being used to sort black plastic flakes in the automotive industry [22]. This system can sort black flakes but is not designed to handle whole packaging items such as black pots, tubs or trays (Section 5.4).

Raman spectroscopy based-technology is, therefore, a potential future method for sorting black post-consumer plastic packaging subject to a commercially viable detection system for the industry becoming available.

5.3.2.3 Laser-induced breakdown spectroscopy/laser- induced plasma spectroscopy

Laser-induced breakdown spectroscopy (LIBS)/laser-induced plasma spectroscopy (LIPS) could be suitable for in-line, real-time detection because it has a rapid response time and can detect spectra remotely. The sampling time for the generated emissions from the laser-heated sample is very short because plasma is generated within nanoseconds of the laser pulse. The emissions are collected, analysed, and then the detection channel is cleared for the next sample. Identification and calculation time are believed to be short, but the overall response time is not known because no commercial systems are available. Pigmentation of the product will not stop the production of the plasma or interfere with the emission detection. Hence, black articles could be identified as well as transparent or light-coloured articles.

A potential limitation is that any modification of the surface, such as surface coatings and labels, will probably affect the results, because only the sample surface is measured.

LIBS and LIPS are not yet commercial technologies for sorting plastics, but they show some promise in this application [23]. This technology could compete with other detection systems in the future if the capability in this area becomes proven and commercial systems are economically competitive.

5.3.2.4 Photoacoustic spectroscopy

Photoacoustic spectroscopy involves measuring the acoustic emission after laser irradiation. The laser radiation is pulsed at an audible frequency, resulting in a sound proportional to the intensity of radiation absorption [24]. The colour and surface condition of the article will not affect the photoacoustic response. It has been tested in recycling plants for automotive and electronic waste sorting [25]. It is not yet known if the robustness and response time of the technique is suitable for use in recycling plants for post-consumer packaging. This would need to be assessed and so this technique is a long way from becoming a competitor technology in this area.

5.3.3 Novel black pigments that are near-infrared detectable

A black pigment can result from a combination of several pigments that collectively absorb all colours. If appropriate proportions of three primary pigments are mixed, the resulting colour reflects so little light as to be called 'black'. Black can, therefore, be described as a lack of all colours of light, or a combination of multiple colours of pigment [8]. It may be possible to select black colourants that absorb in the visible region, but reflect or transmit NIR radiation, allowing for standard NIR detectors in automated sorting systems to record a useful spectrum from a black packaging article. One drawback from moving away from the standard carbon-black technology used in industry (Section 5.3.1) is that the high-IR absorptive properties of this pigment benefits the thermoforming of trays from plastic sheet in that the plastic heats up faster, which increases the operating speed. This is a commercial consideration that will need addressing to assist the industrial take-up of non-carbon black technologies. The technologies that surround the two types of alternative colourants to carbon black (i.e., those that are NIR transparent or NIR reflective) are discussed below.

NIR transparent black colourants are used for laser welding plastic components [8]. They function by allowing IR radiation to penetrate through one component to reach the interface with another component that contains IR absorbers. The result is that the IR radiation causes local heating at the interface of the two layers, causing them to be welded together. Examples of black colourants that are transparent to NIR are the colourants marketed by BASF (e.g., Lumogen Black FK4280 and Lumogen Black FK4281). These products are organic perylene colourants [26] that do not absorb strongly in the IR region, but can be produced with high opacity and blackness. Other examples are: Fuji Xerox colourants with high- black colouration and NIR transparency for photocopy toner; and Treffert GmbH specialised masterbatches for laser welding based on soluble dyes [8].

The BASF literature on Lumogen colourants [8] shows significant transmission at wavelengths longer than 1,000 nm – almost halfway into the NIR spectrum (700–1,400 nm). The Fuji Xerox patent [27] suggests that their pigments, which are based on a squarylium compound combined with phthalocyanine blue colourant, could also be used for laser welding of plastics and Glaser [28] has mentioned in a paper presented to the Joining Plastics conference that they have laser-absorbing and laser-transparent types.

Black colourants that are NIR reflective have been used in coatings, paints and resins that act to reduce solar thermal gain in dark-coloured buildings or automotive products by being low in absorption or reflective of IR radiation. Specific examples include paints and coatings that have high IR reflectivity or low absorption used for applications such as coatings for roof tiles that enable dark colours to be used with less heat gain than achieved usually, and housing for lighting systems to reduce heat build-up. The black colourants used in these types of products may prove to be suitable for black colouration in plastic packaging, yet enable detection by NIR spectroscopy. However, there are already several commercial examples of NIR reflective black colourants that can be used in plastics, such as Sicopal Black K 0095 from BASF and Black 10P922 from Shepherd Colour Company. An IR reflective colourant at low concentration in the polymer matrix could enable a useful IR spectrum to be obtained from the product in reflectance mode.

WRAP in the UK have commissioned work to evaluate a new colourant technology that would enable black items to be separated effectively [29]. These colourants have been found to work in a range of plastics, including APET, CPET, PP, HDPE, PS and PVC. After carrying out work to optimise tint strength, the colourants selected for large-scale trials were Colour tone IRR Black 95530 for PP and CPET and Colour Matrix Dye Black 5 for APET and CPET. Subsequent research in large-scale trials showed that virtually all the PP and PET trays were identified correctly by the NIR detectors and sorted successfully. Figure 5.8 shows how the new black pigment has enabled black trays to be detected and sorted along with the jazz fraction of the plastic recycling stream. When adopting new technology, one of the most important criteria is economic viability and, with this in mind,

Figure 5.8: Coloured jazz fraction including the sorted black trays. Reproduced with permission from the Waste and Resources Action Programme (WRAP), Banbury, UK. ©WRAP.

WRAP carried out a cost assessment to provide a preliminary evaluation of the financial viability of the alternative black pigments, which are relatively expensive compared with conventional carbon-black pigments.

5.3.4 Physical techniques that do not involve spectroscopy-based methods

Both of the physical techniques described below have showed some promise in the sorting of mixed plastic, but both require further development to achieve their full potential.

5.3.4.1 Triboelectric charging

Triboelectric charging of a packaging item is achieved by rubbing it against a charging material and the resulting charge measured using a probe (Section 5.2.2.2). The system is claimed to be suitable for automated sorting but, in its simplest form, is capable of achieving only a binary separation between two polymers with a positive and negative charge relative to a reference material. To achieve effective sorting of a mixture of polymers, a system would have to be set up that uses a series of reference probes. This could be carried out using a conveyor-belt design, but the humidity of the environment that the system operates in would have to be controlled and it can take ≤ 10 s to differentiate between some plastics, which could restrict throughput. Other potential problems with the technique include the adverse influence of surface contamination and surface coatings and labels. Although this type of system has potential, a commercially available sorter that operated using triboelectric principals was trialled in the separation of mixed plastic flake in a study funded by the UK WRAP in 2008 [30], but did not perform well.

Despite these setbacks, research and development is ongoing in this area. A prototype of a new triboelectrical device for the separation of mixed plastic waste streams has been constructed by researchers in Poland [31]. The results that the group obtained enabled them to recommend the device be used for processing mixed plastic.

5.3.4.2 Tack separation

Tack separation works by heating the plastic items by IR radiation. Different plastics become tacky at different temperatures as they approach their meting point. This effect can enable plastics with a low melting point to be separated by their adherence to a conveyor or drum, and has been demonstrated by a US automotive recycling company. One of the problems with black plastic items, however, is that it is more effective to control the heating rate in these processes by using IR radiation that is tuned to the main NIR absorption band of each plastic item, and this band would be masked by the black pigment.

5.3.5 Use of marker compounds

The three principal technologies associated with the use of markers to assist in the sorting of black plastic articles are introduced below.

5.3.5.1 Fluorescent additives

Molecular markers of this type can be co-polymerised into polymers such as PET. They fluoresce in the NIR region if illuminated with a laser and are selected to ensure that they do not affect colour or clarity by absorbing visible light. To covert this theoretical solution effectively into a practical solution, international agreement between plastics suppliers would need to be achieved to ensure global consistency. A more practical approach would be to compound the markers into plastic at levels that they would be detectable using optical sensors (i.e., 0.5 to 20 ppm). Combinations of four fluorescent markers have been shown to enable automated sorting systems to operate effectively. The problems encountered were that colourants in plastics reduced the fluorescent signal strength and carbon-black colourants were found to render the marker system impotent at these levels of addition [32]. Further research, therefore, is required to establish the levels of addition needed to enable effective separations to be achieved. Another possibility is to incorporate these additives into a surface coating. However, the additional cost, food contact compliance, and the need to obtain the agreement of packaging suppliers are potential problems.

5.3.5.2 Taggants

Taggants are a more sophisticated application of fluorescent additives. They are, in general, in the form of fine particles with a combination of colours that encode the identification information. A specific response results if they are exposed to particular frequencies of radiation and they are used for brand verification, anti-counterfeiting and traceability. Technologies suitable for use in plastics have been developed by several companies in the past decade (e.g., 3M Light Reveal Authenticity) and a Kodak Traceless system [33].

As with fluorescent additives, the effectiveness of taggants will be reduced by absorption by carbon black, but may be effective if present in a label or coating.

5.3.5.3 Radio-frequency identification

Radio-frequency Identification (RFID) tags are used widely for tracking plastic pallets, mobile garbage bins and re-useable crates. Passive RFID tags derive power from incident radio-frequency (RF) radiation and do not require a battery. They emit an RFID signal if they detect RF radiation of the correct frequency and intensity. The cost and size of RFID tags have fallen to the point where the tagging of plastic bottles and pallet shrink-wrap for recycling purposes has been proposed [34].

An advantage of this system is that carbon-black colourants would not absorb RF radiation very much. One of the main disadvantages is that they could contaminate the plastic waste stream because they can contain ≈40% metals as well as thermoset plastics. Their removal is possible (e.g., use of a soluble adhesive to aid washing off) but would add to the complexity of the recycling process.

One development in this area that may present a solution for plastic packaging in the future would be the use of lower-cost 'chipless' RFID manufactured from electrically conductive polymers that could be printed onto the polymer [35].

5.4 Other developments in detection and sorting systems

New automatic sorting systems that can boost the recycled content of plastic food packaging are being developed by Axion Consulting and could be available within a few years [36]. Such systems are needed because there are difficulties with plastics available in food and non-food-grades because EU food contact regulations permit only material that was originally food-grade being recycled into new food contact articles. Hence, sorting by polymer type alone is not sufficient and it would not be feasible to determine the complete composition of a plastic item as it passes through a recycling stream to ensure that it complies with the substances in the 'Union List' in EU Plastics Regulation (EU) 10/2011 (Chapter 4). However, if systems were available that could differentiate between food and non-food packaging it would represent a major breakthrough. For example, in the case of PP, this would enable ≈180,000 tonnes/year of waste pots, tubs and trays to be recycled in the UK. Axion Consulting are tackling the problem by carrying out research to develop an automated process that uses diffraction gratings to identify and separate food-grade PP packaging from non-food-grade PP packaging. The process is stated as involving marking food-grade PP packaging with lines (a diffraction grating) that can be scanned by a laser to reflect a specific pattern. Axion Consulting consider that the system could be used for other types of polymer packaging (e.g., PET) in the future, though this situation is not as acute for PET, for which food-grade products are often easier to identify at the collection stage and a significant quantity of PET not used for food contact applications (e.g., washing up liquid) can still be food-grade.

New and improved variants of detectors continue to be developed. An example is the new, high-speed Raman spectroscopy detection system developed as a result of collaboration between Saimu Corporation and Kinki University [37]. This system can attain a purity level of 95% from shredded plastic that is passing by the detector at a throughput of 200–600 kg/h.

Another relatively new system is the POWERSORT 200 sensor system, which has been brought onto the market by Unisensor [38]. This system works using a high-speed laser and, by the use of the opto-electronic spectrum of materials, it can separate PET flake in waste bottle streams from PET flake that contains, or is mixed with, additives and polymers such as Nylon 6,6, PVC, or which has a silicone coating.

The technical possibilities offered by electrostatic sorting have been referred to in Table 5.2 and a study investigating the practicality of the technique has been carried out in Ontario, Canada [39]. The conclusion was that the electronic separation process is not suitable for installation in a typical municipal materials recovery facilities (MRF) because the complexity of operating a shredder and wash line coupled with a float/sink tank and a drying facility feeding an electrostatic separator is probably beyond the capabilities of most MRF.

One innovative technology that has been developed and marketed recently to target specifically the separation of complex polymer systems is the Polymag Process patented in 2005 by Eriez Magnetic [40]. This process is based on the principle of magnetic separation and can separate and recover mixed plastics generated by multiple- material injection-moulding, over-moulding, profile co-extrusion and multiple-layer blow-moulding. To function, a magnetic additive must be incorporated into one of the plastics in the multiple-material product. Then, the shredded material is passed over the high-powered PolyMag Rare Earth Magnetic Roll that effectively separates the component containing the magnetic additive from the component that does not. Trials showed that this system can separate a thin layer of ethylene vinyl alcohol barrier resin from a multiple-layer HDPE blow-moulded article.

The separation and sorting of plastics films is another area in which problems can arise. It was reported in 2011 [41] that the scope and tonnages of plastics being collected from domestic kerbsides had increased and that this had resulted in a higher proportion of film, which posed a challenge to older MRF to maintain efficient separations. In this article, Axion Engineering, a division of Axion Recycling, was described as being 'impressed' by the performance of the air-separation products from German company Nestro Lufftechnik, which can separate light fractions from bulky waste streams, such as films from rigid packaging. Also, Grundon Waste Management were said to have included seven Binder Redwave NIR automatic sorters in the rebuilding of the Colnbrook MRF in Berkshirek and that these could be calibrated to identify and extract plastics, glass, paper and cardboard from the waste stream, even removing difficult-to-handle materials such as Tetrapak cartons and plastic film.

References

1. B. Juijsterburg and H. Goossens, *Resources, Conservation and Recycling*, 2014, **85**, 1, 88.
2. A. Clark, *Plastics & Rubber Weekly*, 2011, 30th September, 11.

3. *Improving Food Grade rPET Quality for use in UK Packaging*, Final Report, Waste and Resources Action Programme (WRAP), Banbury, UK, July 2013.

4. E. Kluenker, *Kunststoffe International*, 2012, **102**, 10, 74.

5. S. Schuh in *Proceedings of GPEC 2010 Conference*, Orlando, FL, USA, Ed., Society of Plastics Engineers, Plastics Environmental Division, Lindale, GA, USA, 8–10th March 2010, Reclamation and Supply Session, Paper RS4, p.7.

6. E. Kosior and R. Dvorak in *Proceedings of GPEC 2010 Conference*, Orlando, FL, USA, Ed., Society of Plastics Engineers, Plastics Environmental Division, Lindale, GA, USA, 9–12th March 2008, Paper RS11, p.37.

7. SuperCleanQ EU-funded FP7 Research Project. http://www.supercleanq.eu

8. *Development of NIR Detectable Black Plastic Packaging*, Final Report, Waste and Resources Action Programme (WRAP), Banbury, UK, September 2011.

9. T. Shelley, *Eureka*, 2010, **30**, 11, 33.

10. A. Buan, *Plastics and Rubber Asia*, 2014, **29**, 203, 17.

11. M. Carvalho, C. Ferreira, L.R. Santos and M.C. Paiva, *Polymer Engineering and Science*, 2012, **52**, 1, 157.

12. Anon, *Plastics and Rubber Weekly*, 2012, 31st August, 8.

13. *End Markets for Recycled Detectable Black PET Packaging*, Final Report, Waste and Resources Action Programme (WRAP), Banbury, UK, July 2013.

14. *Packaging Waste Statistics – Statistics Explained*, Europa, Brussels, Belgium. http://www.ec.europa.eu>eurostat>index.php

15. *Separation of Mixed WEEE Plastics*, Final Report, Waste and Resources Action Programme (WRAP), Banbury, UK, October 2009.

16. Anon, *Resources, Conservation and Recycling*, 2013, **78**, 1, 105.

17. J. Murphy in *Additives for Plastics Handbook*, 2nd Edition, Elsevier, The Netherlands, 2001.

18. M.M Fisher in *Plastics and the Environment*, Ed., A. Andrady, Wiley Interscience, Hoboken, NJ, USA, 2003, p.563.

19. N. Eisenreich and T. Rohe in *Encyclopedia of Analytical Chemistry*, Ed., R.A. Meyers, Wiley, New York, NY, USA, 2000.

20. J. Florestan, A. Lachambre, N. Mermilliod, J.C. Boulou and C. Marfisi, *Resources, Conservation and Recycling*, 1994, **10**, 1, 67.

21. EPA EP1052499 19990109634: Method and Device for Raman Spectroscopic Analysis of Black Plastics, Sony International Europe GmbH, Stuttgart, Germany, 14th May 1999.

22. http://www.unisensor.de/

23. M. Boueri, V. Motto-Ros, W-Q. Lei, Q-L. Ma, L-J. Zheng, H-P. Zeng and J. Yu, *Applied Spectroscopy*, 2011, **65**, 3, 307.

24. V.G. Gregoriou, D.C. Clara and E.C. Charles, Jr., in *Applied Polymer Science*: *21st Century*, Pergamon, Oxford, UK, 2000, p709.

25. *FTIR-Photoacoustic Spectroscopy of Solids*, AMES Laboratory, Iowa State University, IA, USA. http://www.etd.ameslab.gov/etd/technologies/projects/pas.html https://www.ameslab.gov/epsci/ftir-photoacoustic-spectroscopy-solids

26. P. Erk, A. Stohr, A. Boehm, W. Kurtz, J. Mizuguchi and B. Sens, inventors; BASF Aktiengesellschaft, assignee; US7416601, 2005.

27. S. Hasegawa, M. Tian, Y. Ito, K. Anazawa, K. Hirokawa, T. Matsubara, K. Horiuchi, T. Miyahara and M. Furuki, inventors; Fuji Xerox Co., Ltd., assignee; US20090081574, 2009.

28. S. Glaser in *Proceedings of the Joining Plastics Conference*, London, UK, 2006, Paper No.23.

29. T. Gaukroger, *British Plastics & Rubber*, 2013, January, 24.

30. *Domestic Mixed Plastics Packaging Waste Management Options*, Final Report, Waste and Resources Action Programme (WRAP), Banbury, UK, June 2008.

31. M. Zenkiewicz, T. Zuk, M. Blaskowski and Z. Szumski, *Przemyst Chemiczny*, 2013, **92**, 2, 279.
32. S.R. Ahmed, *Assembly Automation*, 2000, **200**, 1, 58.
33. C. Swedberg in *Kodak Markets Optical Marker as RFID Alternative*, RFID Journal, Hauppauge, NY, USA, 2008. http://www.rfidjournal.com/articles/view?3995
34. R. Stigall in *Integrating RFID with Plastic Products and Packaging*, Society of the Plastics Industry, Washington, DC, USA, 2006.
    https://www.plasticsindustry.org/files/industry/scitech/rfid/rfid-stigall-062206.pdf
35. Frequently Asked Questions Section, RFID Journal, Hauppauge, NY, USA, 2009. http://www.rfidjournal.com
36. Anon, *British Plastics & Rubber*, 2013, October–November, 34.
37. A. Tsuchida, T. Yoshida, Y. Tsuchida and H. Kawazumi, *Bunseki Kagaku*, 2012, **61**, 12, 1027.
38. W. von Schroeter, *PETplanet Insider*, 2011, **12**, 3, 12.
39. Anon, *British Plastics & Rubber*, 2006, April, 23.
40. *Plastic Additives and Compounding*, 2007, **9**, 3, 36.
41. Anon, *Plastics and Rubber Weekly*, 2011, 19th August, 9.

6 Recycling technologies for polyethylene terephthalate

6.1 Introduction

Once post-consumer polyethylene terephthalate (PET) products have been separated from other plastics by near-infrared spectroscopy (NIR) detectors, and other contaminants (e.g., fabric and metal fragments) have been removed by the use of a range of other separation systems, they will be placed into a granulator to produce flakes that are ≈25 mm in diameter and these then pass through a caustic (i.e., alkali) hot-wash process to remove surface dirt, paper label fragments and label adhesive. To ensure the highest possible level of decontamination, PET flakes can then go through a further set of decontamination processes and then a chemical cleaning process to remove contaminants that have been absorbed in service. PET flake that has been through such a sequence is then usually labelled as 'high-quality washed flake' (HQWF) and can be sold as a commercial product (e.g., to make PET strapping products) (Chapters 5 and 9).

To enable food-grade products to be made the HQWF must be purified to a sufficiently high level by using one of the many commercial processes available (e.g., Vacurema, United Resource Recovery Corporation, Starlinger). These processes have been shown by the use of 'challenge tests' (Section 4.2.6) to be capable of decontaminating the PET to the extent that the material will meet the requirements of the US Food and Drug Administration (FDA), European Food Safety Authority (EFSA) and the European Union (EU) Plastics Recycling Regulation, (EC) 282/2008. Once the process has received a 'letter of no objection' from the FDA or a 'scientific opinion' from EFSA in Europe, running it within an acceptable quality-control regimen [e.g.,

International Organization for Standardization (ISO) standard, ISO 9001] is usually sufficient to meet the good manufacturing practice requirements of EU Regulation (EC) 282/2008 (Section 4.2.4.2).

A general summary of the various processes that PET HQWF goes through when it is subjected to a 'super-clean' mechanical recycling process (Section 6.2) to generate food-grade recyclate is:

- The initial stage involves drying the HQWF.
- A caustic wash is then applied to the HQWF and it is heated to dry the alkali onto the surface of the flakes to depolymerise it.
- Another heating stage takes place to complete the 'stripping' of the surface of the flakes and also drives off any volatile substances trapped within them.
- A rapid-wash stage to remove soluble substances, such as monoethylene glycol, which will have been generated by the alkali and heat-treatment stages.
- Neutralisation step and rehydration with water to impart some 'plasticisation' to the flakes.

https://doi.org/10.1515/9783110640304-006

- Solid-state polymerisation (SSP) stage to recover molecular weight (MW) (possibly with monomers added).
- Extrusion into granules for bagging, weighing and transportation to processors.

A recent version of the EFSA register [1] of valid applications for authorisation of recycling processes to produce recycled plastics materials and articles intended to come into contact with food submitted under Article 13 of Regulation (EC) 282/2008 showed that of the 89 individual processes present on the list, 81 were for the recycling of PET. This finding shows the dominance that PET has in the food-grade recycling market and this is due to several factors: food-grade PET products are a relatively easily identifiable and isolatable waste stream; they contain relatively low levels of contamination due to the polymers low permeability: its relatively high value. An assessment was made in 2010 by Fraunhofer IVV of the available 'super-clean' PET recycling capacity in Europe [2] and their results are shown in Table 2.12.

The term 'mechanical recycling process' has been used above. When placing different PET recycling technologies into categories, for the sake of convenience, authors choose to use two specific terms:
1. Mechanical processes
2. Chemical processes

Although, as with any categorisation system, there are inevitably 'grey' areas, when this approach is used for PET recycling processes, mechanical processes can be described as those that involve a sequence of steps, such as those shown above. That is, the post-consumer PET flakes are washed to remove contaminants, decontaminated in the melt-state under a high vacuum, and then put through a SSP step to return the PET to a high-MW, decontaminated product that is suitable for added value uses, such as food contact applications. Mechanical processes are described in Section 6.2.

Chemical processes, which are covered in detail in Section 6.3, can be defined simplistically as those in which the post-consumer PET is broken down using routes such as methanolysis or glycolysis by the use of reactive agents and/or catalysts into low-MW substances (e.g., monomers and oligomers). These low-MW substances are then polymerised back into high-MW PET, after purification processes have taken place to remove all contaminants.

Several reviews have outlined the environmental problems that result from the poor control of plastic waste and the resultant need for recycling PET and other synthetic polymers. These reviews often go on to describe the recycling options available and the waste handling and management infrastructure that needs to be put into place to solve the problem. Sinha and co-workers [3] published a review that has a particular focus on the recycling of PET and provides an overview of the chemical recycling processes available to recycle it back into useful products.

6.2 Mechanical recycling processes

This section deals with mechanical processes and these are the most commonly used for the commercial recycling of PET and other commercially important plastics, such as high-density polyethylene (HDPE). Many of the food-grade PET submissions made to the EFSA (Chapter 4) are from users of this type of technology. Also, most of the processes that have passed the EFSA challenge test, and which are currently used commercially to produce food grade-recycled polyethylene terephthalate (rPET), are mechanical-type processes.

These processes are usually set up on the basis that the PET waste stream contains a low (e.g., <5%) amount of non-food PET (Section 4.2.6). Any other polymers and materials are removed by up-stream sorting and separation stages. Then, the PET is converted into flakes and these are purified by a sequence of procedures and techniques before being dried. They are then converted into a melt by heat and shear forces, organic contaminants removed (usually using high temperatures and a high vacuum) and then converted into granules or made directly into product (e. g., sheet for thermoforming) (Section 6.1). However, there are mechanical recycling processes that do always follow this sequence and can be more or less complex, depending on the developments that have taken place, as demonstrated by the examples provided below.

An example of one of the mechanical processes that follows a simpler route to the one provided above and that is already available commercially is the Multi Rotation System (MRS) from Gneuss [4]. This is an extrusion-based system for producing food-grade PET directly from flakes that have not had any pre-treatment. It consists of a multi-rotation devolatilising extruder together with a melt-filtration system with integrated self-cleaning to assist in purification.

It is claimed that its rugged and simple design make it suited to the processing of highly contaminated material [5]. It is also claimed to be capable of enabling the manufacture, from ≤100% rPET bottle flakes, of sheet and food containers of all types for approved direct food contact. In addition, Thiele [6] described how it can be used for a broad range of recycling applications, such as converting fibre or bottle flake waste streams to spinnable pellets, or directly into staple fibre (bulked continuous filament or spunbound). In addition to these technical capabilities, it was reported in the *High Performance Plastics* journal in 2010 [7] that the Gneuss MRS system has received regulatory approval from the FDA for the production of rPET pellets from 100% bottle flake that could be used for food packaging. The FDA approval was specifically for the use of the rPET pellets for the production of hot- and cold-filled PET bottles. The regulatory system in the US for the use of rPET in food contact articles and materials is covered in Chapter 4.

Manufacturers of recycling equipment are looking continually to increase the flexibility of the systems they are offering to the marketplace. An example of this has been reported in an article in *PETplanet Insider* and concerns the manufacturer

Erema, which specialises in the development, manufacture and distribution of plastic recycling systems and technologies [8]. It has developed what it describes as a 'flexible plant concept' for its established Vacurema Prime technology that allows the user to choose from three operating modes according to requirements, and so generate processed rPET in three physical forms. The plant operators can, therefore, adjust the output of the plant to meet changing market demands. The three modes of operation are:

- Mode 1: Production of 100% rPET pellets
- Mode 2: Production of 100% rPET flakes
- Mode 3: Simultaneous production of rPET pellets and rPET flakes

Some of the features of the Vacurema system have been described in a recent article in the journal *Extrusion* [9]. The technology, which has EFSA approval, is claimed to employ a highly efficient, food contact-compliant decontamination step before extrusion and a large-area, ultrafine melt-filtration system. Also, the article states that, when it was tested by an independent testing institute through automatic operation with a food contact control (FCC) feature, it was shown to have a very good energy efficiency rating. The maturity of the commercial market for food-grade PET recycling processes is made apparent by the number of different versions that a manufacture offers for a particular technology. For example, Vacurema Technology is available in several variants, to match the client's specific requirements, for example:

1. Vacurema Basic
2. Vacurema Advanced
3. Vacurema Prime
4. Vacurema MPR

The Vacurema Prime system is one of the processes present on the EFSA recycled food contact plastics register.

Ayodhya and Limaye [10] have reported the results of a study on the mechanical recycling of PET. In the paper they state that the ultimate colour and physical properties of the rPET are dependent on the level of contamination present in the feedstock and flakes. As a consequence, research was carried out by the workers to improve the efficiency of the recycling process and to improve the colour and properties of the rPET that resulted so that they were nearer to that of virgin polyethylene terephthalate (vPET). One result of this research was development of a novel proprietary additive aimed specifically at being used in the production of rPET for food contact applications. The feasibility of using this improved rPET for the injection-moulding and blow-moulding of food contact products was also covered in the paper.

Albert and Schnell [11] described a process (known as the 'Ohl process') that can be used for recycling ground post-consumer PET bottles. The process has two steps, a continuous extrusion process and a discontinuous treatment involving decontamination and post-condensation of the PET back into a high-quality product. The tumble reactor employed in the decontamination stage can also be used to purify and carry out post-condensation work on other polyester polymers (e.g., polybutylene terephthalate) and polyamides. Another use for it is in the drying of pellet products in chemical and pharmaceutical industries.

Starlinger has several recycling technologies for bottle-to-bottle PET and high-viscosity applications on the market. The basic modules of this technology are extruders for the mechanical recycling of the PET bottle flakes and other physical forms. These systems range from the budget versions (e.g., recoStar PET FG and recoStar PET FG+), which can be supplemented by an SSP reactor, to the recoStar PET iV+ system for bottle-to-bottle recycling that can achieve very high levels of decontamination, to the recoStar PET iV+ Superior for the most stringent purity requirements. In this latter process, the rPET is decontaminated twice under vacuum (before and after the extruder) to achieve a very high level of decontamination [12]. An article by the company in the *International Fiber Journal* claims that the use of a Starlinger recycling system to produce an rPET that has an intrinsic viscosity similar to that of a vPET resin also provides the potential to use ≤100% rPET to produce final products [13].

The 'Diamet' decontamination recycling technology from Holfeld Plastics Ltd involves an EFSA approved process that mixes 50% rPET with 50% vPET and heats the mixture to a temperature where crystallisation occurs. The resulting product is then decontaminated through two consecutive extrusion steps that include multiple vacuum degassing. Food contact-grade sheets are then extruded from the material and thermoformed into food packaging trays and punnets [14].

An article in *PETplanet Insider* [15] has reported on Extricom's RingExtruder RE system that is stated as having been used for PET bottle-to-bottle recycling since 1998. This system is said to be capable of producing high-quality raw materials that can be used in food contact applications. The RingExtruder RE7 XPV product is claimed to be able to produce ≤4,500 kg/h of rPET.

Pichler and co-workers [16] of Next Generation Recyclingmaschinen GmbH have described the development of a process (called the 'LSP process') said to offer advantages for the recycling of PET. The process is said to work by shifting the polycondensation reaction into the liquid phase of the rPET and then using the greater molecular mobility that exists within this phase to increase the reaction rate. They also reported the results of recycling trials undertaken using the LSP process to recycle post-consumer PET fibre. An increase in intrinsic viscosity was achieved within a few minutes, and the levels of decontamination were significantly better than those required for the production of food-grade rPET.

An article in *PETplanet Insider* [17] described how the M-PET process, which was patented in 2002, can be used to recycle PET bottles without needing a SSP stage. The method that the process uses was developed after research had been conducted on the interactions that took place between different polymer molecules in contact with polyesters. This work discovered that silane compounds (preferably polyhydrosiloxane in combination with a plasticiser such as di-octyl phthalate) can create compounds with polyester molecules. By this route, as low-MW oligomers and molecules are generated within the PET due to its degradation during, for example, processing, they are re-combined and long-chain molecules created that can have a higher MW than some of the original PET molecules. In addition to these re-combination and chain-extension reactions, crosslinks can also be introduced into the PET, which counteracts its tendency to crystallise. It was reported in the article that the M-PET process has been scaled up to an industrial scale using a 75-mm twin-screw extruder with co-rotating parallel screws.

One of the problems that can be encountered if PET is recycled is a loss of MW due to chain scission and this in turn can lead to a loss of important physical properties which limits its end-use applications (Section 9.15). A Japanese group [18] looked at how the use of radio waves can lead to an improvement in the intrinsic viscosity (i.e., MW) of rPET. Their study involved use of an industrial radio-frequency heating process during a thermal, solid-phase polymerisation recycling method. The results obtained indicated an increase in the intrinsic value of rPET.

In an article in the *International Fiber Journal* [19], Sikoplast describe their KTE system for recycling PET bottle flakes, film and fibre. The system is said to include a crystallisation-dryer unit inserted before the extruder and a special silo suitable for drying free-flowing and non-free-flowing materials. The process is claimed to eliminate the loss of intrinsic viscosity in the rPET and to be characterised by low consumption of specific energy.

Torkelson and co-workers [20] discussed the use of solid-state shear pulverisation (SSSP) for the use of recycling PET. They described how the SSSP process can lead to *in situ* mechanochemistry and enhanced dispersion relative to melt-state processes. In the case of PET, they claim that using SSSP overcomes the problems of recycling the material for high-value applications, due to a loss of MW, because it results in low levels of branching and enhanced dispersion of heterogeneous nuclei leading to increased melt viscosity and crystallisability.

Flake-to-Resin (FTR) is the proprietary recycling technology of Uhde Inventa Fischer for producing food-grade resin capable of being made into food-grade rPET packaging [21]. The FTR process consists of two steps. In the first step, impurities are eliminated from the post-consumer PET and a resin produced. In the second step, the partially depolymerised rPET resin is blended with vPET prepolymer during the polymerisation process and the mixture is polymerised to a bottle-grade rPET. In this way, the FTR process can replace ≤50% of PET production feedstocks.

Welle [22] carried out an investigation into the decontamination kinetics of a super-clean PET bottle recycling process based on solid-state polycondensation. This work involved calculation of diffusion coefficients from decontamination kinetics simulated using migration models for spherical pellets. Data showed that the decontamination of rPET pellets obeyed Fickian laws and that the diffusion coefficients were not influenced by process conditions involving a vacuum or inert gas. Results also showed that the diffusion equations were suitable for simulating the efficiency of the decontamination that takes place during the recycling process.

6.3 Chemical recycling processes

As mentioned in Section 6.1, in chemical recycling processes the PET is depolymerised into its monomers [e.g., ethylene glycol (EG), terephthalic acid (TPA) and dimethyl terephthalate], or its oligomers. Three approaches are possible:
- Methanolysis
- Hydrolysis
- Glycolysis

Although not used to the same degree commercially as the mechanical processes, chemical recycling processes are attractive with regard to their potential ability to remove a greater range of contaminants, and to enable the recycling of difficult products, such as multiple-layer films and films containing barrier layers. As a consequence, as shown by the publications reviewed below, these chemical recycling systems are a particularly active area of research and several new catalyst systems/ chemical reagents [e.g., neopentyl glycol (NPG), mesoporous metal oxide spinels, zinc acetate catalyst] for the depolymerisation of post-consumer PET have been reported in the literature over the last few years.

The chemical recycling of PET is increasingly of interest as a way of generating valuable feedstock substances for various chemical processes (e.g., re-polymerisation into polyesters) (Chapter 9). Lopez-Fonseca and co-workers [23] conducted a study whereby PET was subjected to glycolysis using an excess of EG in the presence of different simple chemicals that acted as catalysts. The compounds investigated as catalysts were zinc acetate, sodium carbonate, sodium bicarbonate, sodium sulfate and potassium sulfate. Yields of ≈70% of the monomer *bis*(2-hydroxyethyl)terephthalate (BHET) were obtained with zinc acetate and sodium carbonate acting as catalysts, in the presence of a large excess of glycol, at 196 °C and a PET:catalyst molar ratio of 100:1. Once isolated and purified, the monomer was characterised by a range of analytical techniques: elemental analyses, differential scanning calorimetry (DSC), Fourier-Transform infrared spectroscopy (FTIR), and nuclear magnetic resonance spectroscopy (NMR). Data revealed that although the intrinsic activity of the zinc acetate was significantly higher than that of sodium carbonate, the latter could be used as an effective, eco-friendly catalyst for the glycolysis of PET. The group also reported on the results that had been obtained on an exploratory study carried out to investigate the application of this catalytic technology to recycling complex PET waste streams (i.e., highly coloured and multiple-layered products).

Research work continues to be carried out in this area to increase the range of post-consumer PET products that can be recycled effectively (e.g., laminated films). An example of the research that has been carried out is the EU-funded project SuperCleanQ [24]. This project was funded *via* the European Commissions' Framework 7 programme and ran from 2011 until 2014. One of the features to come out of it was a novel decontamination system which has the potential to be exploited in two areas associated with the recycling of food-grade PET:

1. As part of a 'super-clean' process designed to be used in the recycling of relatively uncontaminated PET, such as bottles, back into food-grade product.
2. As part of a super-clean process targeted at the large amount of PET (700,000 tonnes) that is regarded as 'unrecyclable' back into high-quality, food-grade PET because it is black, blended with other polymers, or contains barrier layers coatings (Section 2.3.1).

In addition to working on a novel decontamination process for recycling complex PET products, SuperCleanQ also assessed the capability of different detector systems to separate black and dark-coloured plastic products (Figure 6.1) from plastic waste streams (Chapter 5). In addition, to assist the recycling industry with its efforts to detect contaminants [e.g., polylactic acid (PLA) and limonene] in rPET, it also developed an offline analytical method that was published as a CEN Technical

Figure 6.1: Examples of black plastic packaging currently destined for landfill or energy recovery. Reproduced with permission from the Waste and Resources Action (WRAP), Programme, Banbury, UK. ©WRAP.

Specification (Chapter 7), and a NIR-based inline monitoring system for processes such as extrusion (Section 6.5.2).

In addition to investigating new chemical reagents and catalysts for the chemical recycling of PET, researchers have also looked into ways in which the process could be carried out using less energy than existing processes, and so are more environmentally friendly and cheaper. As mentioned in Chapter 3, these considerations are important for recycling any material, particularly the economic viability of a particular method so that it has commercial potential. The use of less energy was one of the benefits claimed by Siddiqui and co-workers [25] for their depolymerisation process for PET, which is based on methanolic pyrolysis (i.e., methanolysis) under microwave irradiation. They applied the process to post-consumer PET bottles in a sealed microwave reactor using methanol. The pressure and temperature were controlled and recorded, and the reaction was carried out in the presence of a catalyst (zinc acetate) and also in its absence. Results showed that depolymerisation was achieved more effectively as temperature and microwave power were increased. For example, a high degree of depolymerisation resulted at ≈180 °C and a microwave power rating of 150 W. Examination of the products of the reactions by FTIR and DSC showed that the principal substance that resulted was dimethyl terephthalate. With regard to the reaction time, the group found that most of the degradation of the PET took place within the initial 5–10 min, which they regarded as shorter than conventional pyrolysis methods and hence more energy-efficient.

A team of researchers in Korea [26] investigated the subcritical and supercritical glycolysis of PET using EG to generate the monomer, BHET. The supercritical glycolysis was carried out at 450 °C and 15.3 MPa, whereas the subcritical glycolysis was carried out at 350 °C and at 2.49 MPa, or at 300 °C and 1.1 MPa. Results showed that high yields (90%) of BHET were obtained under subcritical and supercritical conditions. With regard to the optimum reaction time, the group found that for the same PET/EG weight ratio of ≈0.06, it was 30 min for the supercritical process, and 75 min and 120 min for the two versions of the subcritical glycolysis. A range of analytical techniques were employed to characterise the PET and reaction products, including gel permeation chromatography (GPC), reverse-phase high-performance liquid chromatography, DSC, proton NMR and ^{13}C-NMR. The team concluded that the supercritical glycolysis route, with its short reaction time, would be suitable for a chemical recycling process that required a high throughput.

Allen and co-workers [27] reported on the development of a family of organic catalysts that can be used to depolymerise PET. The group used computational chemistry to gain a clear mechanistic understanding of the depolymerisation process and, in the paper, presented to the GPEC 2010 conference, discussed how the new recycling process could be used to recycle PET, particularly in the context of the progress that had been made to date with bottle-to-bottle recycling.

Vanini and co-workers [28] developed a new method for chemical recycling of post-consumer PET bottles. Their method involves depolymerisation in an alkaline solution (7.5 mol−1 sodium hydroxide) at 100 °C using the surfactant cetyltrimethylammonium bromide (CTAB) as a catalyst at 1×10^{-2} mol^{-1}. The presence of the CTAB was found to increase the reaction performance by ≈85%. They also found that if a mixture of sodium hydroxide and CTAB was used in a proportion of 4:1 (%v/v) the reaction time for a 2-g sample of PET was reduced from 6 h to 2 h. The monomeric product that resulted from the depolymerisation (TPA) was characterised by thermogravimetric analysis (TGA), DSC, FTIR and mass spectrometry. A further advantage of the process was that the CTAB catalyst remained in the alkaline, aqueous phase and did not interfere with the process used to purify the TPA.

A group in India investigated optimisation of the PET glycolysis process using a response surface methodological (RSM) approach involving two-component modeling using the glycolysis time and temperature [29]. RSM was used to predict the optimal glycolysis time and temperature for the recycling of PET scrap. A central composite rotatable design for two variables at four levels was chosen as the experimental design. Data obtained from measurement of the properties fitted into a second-order equation and plotted as three-dimensional surface plots using a programme developed in MATLAB v5.

Analysis of variance was used to evaluate the validity of the model. The optimum operating conditions for the glycolysis time and temperature were found to be 6.5 h and 180 °C, respectively, and under these conditions, the hydroxyl value and glycolysis conversion percentage were found to be 38.14 mg KOH/g and 95%, respectively, a desirability level of 97%. At the same desirability level, the acid value of the glycolysis product was found to be 12.2 mg KOH/g and the number-average MW 695 g/mol.

Nigar and co-workers [30] discussed the recycling of PET soft drink bottles using diethylene glycol (DEG) and 4-(dimethylamino)pyridine as a catalyst. The influence on the process of using different amounts of the catalyst was investigated, and the original post-consumer PET and glycolysis products were examined by FTIR. Results showed that the main product of the glycolysis reaction was the monomer and that this contained small amounts of dimer and oligomers. The group also determined the optimum concentration for the catalyst and showed that it was capable of generating a greater yield of products than is normally the case for glycolysis reactions. The results also showed that an excess of catalyst hindered the reaction.

A group from Eastern Europe [31] also studied the use of DEG to recycled post-consumer PET soft drinks bottles by glycolysis. They investigated the influence of zinc acetate as a catalyst on the process and undertook experiments in which the catalyst was present or absent. The products obtained by the different reactions were investigated by viscosity measurements and by functional analysis of glycols as a function of time. Results showed that an equilibrium mixture of ethylene terephthalate monomer and dimer could be achieved at long reaction times

The use of several novel catalysts for recycling PET *via* glycolysis has been reported by Imran and co-workers [32]. The objective of the research was to develop an optimised process using novel catalysts for the production of highly pure BHET monomer from post-consumer PET. To this end, the multiple-nation group synthesised and characterised novel manganese, cobalt- and zinc-based mixed-oxide spinels and then investigated their effectiveness in the glycolysis of post-consumer PET, which comprised mainly soft drinks bottles. The novel catalysts were employed, together with an excess of EG, to depolymerise the PET back to the BHET monomer. The influence of different reaction parameters (temperature, catalyst type, reaction time, EG/PET ratio, and catalyst/PET weight ratio) were investigated. The results revealed that the catalyst that yielded the highest amount of BHET (92.2 mol%) under mild reaction conditions (260 °C and 5 atm) was zinc manganite tetragonal spinel. The group thought that the high catalyst activity of this particular spinel could be due to its greater surface area, the presence of mild and strong acid sites, and its overall higher concentration of acid sites.

In addition to the two glycols mentioned above, NPG has been used for the chemical recycling of rPET by glycolysis. This particular glycol was used by an Indian team from two colleges in Mumbai [33] to depolymerise rPET into low-MW products that could then be used in the manufacture of polyurethane (PU) coatings. The NPG was used at a molar ratio of 1:6 (PET:NPG) and the glycolysis reaction was carried out at 200–220 °C in the presence of 0.5% zinc acetate as a transesterification catalyst. The group used MW determinations and the hydroxyl value of the glycolysed oligomer to follow the progress of the reaction and the oligomer was also characterised using GPC and DSC. The glycolysed monomer that was obtained by the reaction was purified and then, once it had been characterised using a wide range of analytical techniques, it was reacted with adipic acid, isophthalic acid and trimethylol propane to produce a polyester polyol. Then, this polyol was reacted with a range of commercially available isocyanates to produce PU coatings that were applied to mild steel panels and their optical, mechanical, chemical and thermal properties evaluated using a range of techniques.

A research team from Sao Carlos University in Brazil [34] used SSP to examine the thermooxidation and polycondensation reactions of PET. The team carried out the SSP work in the static mode at a low pressure (30 mmHg) for various times and at various temperatures and used the RSM to examine the effect of these different reaction conditions on the intrinsic viscosity of the rPET. Data showed that the effects of temperature on the intrinsic viscosity were less pronounced than expected as a result of the SSP being surface diffusion-controlled. Also, the SSP efficiency for temperatures ≤230 °C and dwell times of >330 min showed a lower increment in the intrinsic viscosity than occurred at 215–230 °C. The group suggested that this effect was a consequence of a higher conversion rate for the side reactions with increasing temperatures, their cumulative effect with increasing dwell time, and a lower increment of the polycondensation conversion rate with temperature under a less severe vacuum. They also reported that the processing temperatures and the intervals required for recycling PET into new bottles were moderate.

Tawfik and Eskander [35] investigated the use of ethanolamine for the aminolytic depolymerisation of PET in the presence of dibutyl tin oxide acting as the catalyst. The reaction was carried out at 190 °C under atmospheric pressure, and the white precipitate obtained as the product was examined using a range of spectroscopic [FTIR, NMR, X-ray diffraction and mass spectrometry (MS)], thermal analysis (DSC, differential thermal analysis and TGA) and chemical- characterisation techniques. These tests identified the product as being bis(2-hydroxyethylene)terephthalamide, which could be used as a starting material for the production of PU products, such as adhesives and coatings (Chapter 9).

A group in India [36] studied the use of different amine compounds for the depolymerisation of waste PET at an ambient temperature and pressure. The products of the various amine depolymerisation reactions were analysed by FTIR, NMR and DSC and the results obtained showed that they were the corresponding n-alkyl terephthalamide. The data also demonstrated that the depolymerisation process had been effective and successful.

An example of a new chemical method for the recycling of PET is a patented technology developed by Ioniqa Technologies. This process uses a depolymerisation route to convert coloured and low-grade PET waste into a colourless monomeric product which can be used to manufacture high-value PET products such as food-grade bottles. PET products made by this route are complaint with the EU Plastics Recycling Regulation (EC) 282/2008 due to the purifying nature of the process. The operating costs of the Ioniqa Technologies process are low due to the use of low temperatures and because the smart magnetic fluids catalyst can be re-used upon application of a magnetic field.

Several other groups have published papers in the last couple of years concerning development of new recycling processes for the production of food-grade PET. An example of such a group is a team from Aristotle University in Thessalonika [37]. They carried out the chemical recycling of PET by a depolymerisation process using ethanolamine, with and without catalysts, in a sealed microwave reactor.

The high level of research activity demonstrated in this section shows that funding providers see this as an attractive area within which to develop and market new systems and processes. It also indicates that the degree of competition within this sector will increase in the coming years.

6.4 Removal of contamination from post-consumer polyethylene terephthalate

Contaminants can be removed from post-consumer PET in several ways:
– Washing using substances such as alkalis or proprietary chemical mixtures.

- Filtration of the polymer melt.
- Filtration of the mixture of low-MW substances generated by chemical recycling processes.
- Use of a high vacuum.

The first two of these ways are mentioned in this section and some discussion of the third and fourth options has taken place in Sections 6.2 and 6.3.

6.4.1 Washing of polyethylene terephthalate

As mentioned in Section 6.1, one of the processes that takes place at the start of a PET recycling operation is washing of the post-consumer PET products and flake. Krehula and co-workers [38] investigated the influence that different washing regimens have on the levels of contaminants that remain within PET bottle material. To achieve this goal, samples of post-consumer PET bottles were washed in sodium hydroxide at two temperatures (70 and 75 °C) for two time periods (15 and 30 min) and the cleaning efficiency of the washing processes determined by the identification of residual impurities and degradation products in the PET. Data on these substances was obtained by analysis of samples before and after washing using a range of analytical techniques (gas chromatography–MS, GPC, and DSC). Results revealed only low levels of oligomeric material were present in samples, suggesting that low levels of degradation had occurred, and that high levels of purity had been obtained, particularly at the higher temperature (i.e., 75 °C). This high level of purity was due to the efficient removal of external contamination (e.g., adhesive) and low-MW substances, including those formed as a result of degradation in service due to ageing. Overall, the group concluded that the 15-min wash at 75 °C was the best option because it produced PET that exhibited low levels of degradation and contamination in a short process time.

In several sections of this book, the potential problems that can result due to the presence of PLA in the PET waste stream have been highlighted. Two researchers from Cranfield University in the UK investigated the possibility of using selective chemical recycling to separate mixtures of PLA and PET [39]. They evaluated the glycolysis of post-consumer PET using three catalysts: zinc acetate, zinc stearate and zinc sulfate. Data revealed that zinc acetate was the most soluble and effective of the three catalysts, so this compound was used to investigate chemical recycling by solvolysis of a mixture of PLA and PET in methanol or ethanol. Under a particular set of conditions the catalyst was effective at depolymerising the PLA into lactate esters, but the PET was left as an unconverted solid. The researchers considered that their findings provided a strategy to selectively recycle mixed PLA/PET plastic waste by converting one plastic to a liquid and recovering the other, un-reacted one, by filtration.

6.4.2 Melt filtration

One of the ways to remove contamination in post-consumer rPET during processing is by filtering its melt stream. This can be done by fitting an inline melt-filtration system to an extruder which, by a combination of heat and shear forces, converts the rPET flakes into a free-flowing polymer melt. In the filter, the rPET melt passes through a mesh or perforated cylinder with holes of a size that can be chosen depending on the characteristics of the impurities that are expected. The non-melted impurities are retained on the filter and the filter mesh or cartridge is removed readily for cleaning or replacement.

One of the companies involved in the development of melt-filtration technology for many years is the German firm Kreyenborg. They are credited with developing the first backflush filter in 1988, and it is still very prominent in the melt filtration of rPET process lines for the manufacture of food packaging films, packing straps and bottle resins. It was announced in 2011 that Kreyenborg had delivered and installed the first commercial line in which rPET flakes could be fed into a main stream of vPET to a customer in North America [40]. A key component in this process was reported as being a Kreyenborg-patented Vtype backflush filter.

Shah of Pall Corporation [41] described how the Chung Shing Textile Company used Pall's advanced Continuous Polymer Filter melt-filtration technology to produce high-quality, sustainable polyester fibre, such as GreenPlus fibre, from post-consumer PET bottles. It was also mentioned that some of the uniforms worn by athletes at the 2012 London Olympic Games were made using GreenPlus fibres.

Wrobel and Bagsik [42] published the results of a study they undertook to show the importance of melt filtration for the removal of impurities in rPET. They evaluated the influence of several filtration variables (e.g., size of the grid filter) on the physical properties of rPET products, such as films. Results confirmed that the approach used to filter impurities from the PET melt could have an influence on the mechanical properties of the recycled product as a result of its effectiveness in removing impurities. They also found that better results could be obtained if a filtration system was used in an extrusion process capable of automatic filter changes.

Backflush filters are characterised by two functions: filtration and screen cleaning. Their function is to remove impurities and dirt particles from the polymer melt and to provide a constantly clean filtration area due to automatic screen cleaning. Radig [43] of Maag Pump Systems AG described commercial development of a melt-filtration system that uses constant-volume backflush screen changers for the fully automated filtration of polymer melts and screen cleaning, together with a high-process throughput. According to Radig, the system has a dead zone-free channel design, which is optimised for the direct recycling of various polymers, including the recycling of PET bottle flakes to produce deep-drawing films. The process can be retrofitted into existing extrusion lines as well as being integrated into new installations according to the manufacturer.

One of the innovations that resulted from the SuperCleanQ project [44] was a novel decontamination process comprising an innovative degassing/filter-system for the decontamination of the rPET in the melt state. Within this scheme, a high-efficiency purification process was established which required only a low quantity of a stripping agent, such as carbon dioxide or water. Also, several new filter technologies were evaluated during the project. The attributes of the SuperCleanQ decontamination system, which is intended to be used in mechanical recycling processes, include:

– It is easy to integrate into existing food-grade PET recycling lines to enhance their purification efficiency.
– It requires only relatively small amounts of stripping agents that have a low toxicity.
– Results of volatile organic carbon analyses obtained on the purified PET showed that that it achieved very good purification.
– No evidence has been obtained to suggest that it reduces the MW of the PET (i.e., depolymerisation-type degradation does not take place).

6.5 Quality control of recycling processes

Several areas within a PET recycling process require quality control. These areas, particularly with regard to food-grade rPET, are addressed in the EU Regulation (EU) 282/2008 (Chapter 4).

Two specific areas which will be covered here because they have been the subjects of research activity in recent years are the control of low-MW chemical contamination to ensure compliance with food contact regulations, and the monitoring of the quality of the rPET in the melt-phase at the end of the process. The types of systems and techniques (i.e., chemical analysis and inline monitoring) that can be used for these purposes in recycling systems can also be used to control the quality of rPET products, and so they are also included in Chapter 7.

Such techniques can be included in what can be described as industry 'process analytical technology' (PAT) tools for real or near-real-time measurements of chemical composition and/or physical properties. Such measurements can also yield data for control and optimisation of material properties and processing conditions. PAT differentiates between inline, online and offline measurements:

– *Inline* measurements are implemented directly within the processing line, resulting in very short (or even non-existent) delays for sampling. However, inline sensors may interfere with the main process and the sensors can be influenced by temperature or pressure.
– *Online* techniques require a sampling stream to be diverted from the process flow line and transferred to the measurement device. The sampling stream is,

therefore, isolated from the main stream and delays might occur due to material storage in the bypass system.

- *Offline* techniques*, which still dominate process analyses, are usually relatively expensive and lead to large delays between the occurrence of defects or process instabilities and their detection.

*The SuperCleanQ CEN Technical Specification analytical method for the detection and quantification of six nominated marker compounds (Chapter 7) and the analysis work required for EFSA and FDA challenge tests are examples of offline techniques.

6.5.1 Quality control of low-molecular weight contaminants

The presence of certain organic compound contaminants in recycled food-grade PET can cause regulatory problems. For example, acetaldehyde is created within the PET when it undergoes degradation during the lifetime of the original product, or during the processing or recycling of the polymer. This substance has a specific migration limit into food in the EU Plastics Regulation (EU) 10/2011 and testing is carried out on PET food packaging to ensure that this limit is not exceeded. In addition to any health and safety concerns, acetaldehyde migration into food can affect its organoleptic properties because only 10–20 parts per billion of acetaldehyde can result in an off-taste. For these reasons, this compound has been included in the list of six compounds detected and quantified by the analytical method developed by the research project SuperCleanQ (Chapter 7). The presence of other contaminants such as limonene and methyl salicylate can indicate that there is a problem with the recycling process because these constituents are present in the food products that the PET packaging has contacted during its lifetime, and they should have been removed.

The existence of the challenge test in the EU and the US has provided regulators with confidence that recycling processes for PET and a few other polymers (e.g., HDPE) are capable of removing contaminants from post-consumer PET products (particularly bottles) to a level that present a negligible risk to consumers. This ability to check a process and to receive regulatory 'approval' *via* a 'letter of no objection' (FDA) and 'scientific opinion' (EFSA) has provided industry with the confidence to invest in and develop several recycling processes and, as this section has shown, many of them are now used commercially. Challenge tests are covered in more detail in Chapters 4 and 7.

6.5.2 Quality control of the recycled polyethylene terephthalate in the melt-state

Recycling processes usually generate rPET pellet or a specific rPET product (e.g., film) and it can be possible to install systems to monitor the quality of these products. The techniques described in this section can also be used during the manufacturing of rPET products by processes such as injection moulding or extrusion.

The real-time inline/online monitoring of polymers in the melt has been of increasing importance and interest to the polymer industry for several years because of their advantages over offline techniques. A wide range of process-monitoring techniques are available to generate accurate rheological and compositional data for assessment of real-time processes [45]:

1. *Process rheometry* uses measurement of variations in pressure and temperature to derive shear and extensional characterisation.
2. *Visualisation* uses stress birefringence, particle streak velocimetry and laser sheet lighting to carry out work that would include the illumination of planes across die flows and derivation of polymer melt shear and extensional rheology.
3. *Ultrasound* – high resolution (e.g., nanosecond) measurements of transit time have been used at a frequency of ≈20 Hz to monitor properties such as bulk melt temperature in the extrusion of polymer melts.
4. *Process modelling* – finite element analysis is used with process measurement techniques to improve process understanding and modelling.
5. *Spectroscopy* – there are a range of options in this class:
   - Mid-infrared spectroscopy (MIR)
   - NIR spectroscopy
   - Ultraviolet spectroscopy
   - Raman spectroscopy

The spectroscopy group of techniques can provide information on chemical composition and molecular information. Of these techniques, MIR spectroscopy is more troublesome due to the problems associated with the need for small path lengths to avoid information overload and the absence of robust and relatively cheap optical-fibre technology. The other spectroscopy techniques tend not to suffer from the same drawbacks, and so they tend to be used in industrial environments.

The inline NIR system developed by the SuperCleanQ project (Chapter 7) belongs to the spectroscopy group. The SuperCleanQ project developed a system that used NIR to detect biodegradable contaminants in the two food-grade plastics for which recycling is practiced the most frequently: PET and HDPE. Two specific contaminants were targeted:

1. PLA
2. Oxobiodegradable additives (e.g., the stearate carrier)

Using multivariate analysis and partial least-square analysis, chemometric models in the form of calibration curves were constructed from offline and online calibration data. These showed that the concentration of PLA or oxobiodegradable additives could be predicted by measuring the IR spectrum during injection moulding. For example, PLA could be detected in concentrations as low as 0.01%. Results also showed that the system was very sensitive to changes in the composition of the polymer melt, thereby demonstrating early warning of non-compliance in processed rPET. This NIR inline system can, therefore, detect degradable contaminants in rPET melt-processing streams at concentrations below those likely to cause significant deterioration in the mechanical properties of PET. It is also capable of detecting other contaminants, such as limonene and methyl salicylate and so, in some respects, it could be used in conjunction with the SuperCleanQ chemical-analysis method (Chapter 7).

It can be seen from the list above that, though several techniques are available for this type of work, each individual generic technique is specialised and is employed for a particular purpose. Whether they are used in any particular recycling process would be very dependent on the specific requirements of that process. Though installing an inline monitoring system within a process line (e.g., at the die end of an extrusion line or in the nozzle of an injection-moulding machine) enables real-time quality control and could provide an organisation with the following:

1. Faster response to deviations in process conditions
2. Reduction in manufacturing losses due to non-compliance of products

References

1. *EFSA Register – 15th Update*, European Food Safety Authority, Parma, Italy, 28th January 2014.
2. F. Welle, *Kunststoffe International*, 2011, **101**, 10, 45.
3. V. Sinha, M.R. Patel and J.V. Patel, *Journal of Polymers and the Environment*, 2010, **18**, 1, 8.
4. No Author, *PETplanet Insider*, 2011, **12**, 12, 14.
5. A. Hannemann, *Kunststoffe International*, 2011, **101**, 10, 83.
6. U. Thiele, *Chemical Fibres International*, 2014, **64**, 1, 42.
7. Anon, *High Performance Plastics*, 2010, April, 11.
8. Anon, *PETplanet Insider*, 2011, **12**, 12, 12.
9. Anon, *Extrusion*, 2014, **20**, 3, 26.
10. S.R. Ayodhya and C. Limaye in *Proceedings of the 2012 ANTEC Conference*, Mumbai, India, 6–7th December 2012, Ed., Society of Plastics Engineers, Brookfield, CT, USA, 2013, p.634.
11. D. Albert and H. Schnell, *Kunststoffe International*, 2013, **103**, 1, 16.
12. Anon, *PETplanet insider*, 2011, **12**, 3, 20.
13. Starlinger & Co. GmbH, *International Fiber Journal*, 2011, **25**, 2, 8.
14. *Holfeld Takes Food Safety to a New Level*, Holfeld Plastics Ltd, 20th March 2013. www.holfeldplastics.com/english/newsarchive/holfeld-takes-food-safety-to-a-new-level/
15. Anon, *PETplanet Insider*, 2012, **13**, 3, 12.
16. T. Pichler, K. Brzezowsky, D. Hehenberger and M. Heinzlreiter, *Kunststoffe International*, 2014, **104**, 2, 53.
17. Anon, *PETplanet Insider*, 2012, **13**, 12, 20.

18. M. Ogasahara, M. Shidou, S. Nagata, H. Hamada and L. Yew Wei in *Proceedings of the 68th ANTEC Conference*, Orlando, FL, USA, 16–20th June 2010, Ed., Society of Plastics Engineers, Brookfield, CT, USA, 2010, p.1698.

19. Anon, *International Fiber Journal*, 2010, **25**, 3, 41.

20. J.M. Torkelson, P.J. Brunner, C. Pierre, A. Millard-Swan and K. Pukala in *Proceedings of the 68th ANTEC Conference*, Orlando, FL, USA, 16–20th June 2010, Ed., Society of Plastics Engineers, Brookfield, CT, USA, 2010, p.2221.

21. Anon, *PETplanet Insider*, 2010, **11**, 11, 22.

22. F. Welle, *Packaging Technology and Science*, 2014, **27**, 2, 141.

23. R. Lopez-Fonseca, I. Duque-Ingunza, B. De Rivas, S. Arnaiz and J.I. Gutierrez-Ortiz, *Polymer Degradation and Stability*, 2010, **95**, 6, 1022.

24. Anon, *European Plastics News*, 2012, **39**, 3, 25.

25. M.N. Siddiqui, H.H. Redhwi and D.S. Achilias, *Journal of Analytical and Applied Pyrolysis*, 2012, **98**, 1, 214.

26. M. Imran, K. Bo-Kyung, H. Myungwan, C.B. Gyoo and K.D. Hyun, *Polymer Degradation and Stability*, 2010, **95**, 9, 1686.

27. R.D. Allen, J.L. Hedrick, Fukushima Kazuki, H. Horn and J. Rice in *Proceedings of the GPEC 2010 Conference*, Orlando, FL, USA, 8–10th March 2010, Ed., Society of Plastics Engineers, Plastics Environmental Division, Lindale, GA, USA, S2010, Recycling Session, Paper R17, p.27.

28. G. Vanini, E.V.R. De Castro, Eloi Alves Da Silva Filho and W. Romao, *Polimeros*, 2013, **23**, 3, 425.

29. S. Katoch, V. Sharma, P.P. Kundu and M.B. Bera, *ISRN Polymer Science*, 2012, Paper No.630642, p.9.

30. M. Nigar, A. Feroze, N. Rashid and E.B. Coughlin in *PMSE Preprints Volume 105*, American Chemical Society, Division of Polymeric Materials Science and Engineering, Washington, DC, USA, 2011, p.1096.

31. E. Rusen, A. Mocanu, F. Rizea, A. Diacon, I. Calinescu, L. Mititeanu, D. Dumitrescu and A-M. Popahave, *Materiale Plastice*, 2013, **50**, 2, 130.

32. M. Imran, D.H. Kim, W.A. Al-Masry, A. Mahmood, A. Hassan, S. Haider and S.M. Ramay, *Polymer Degradation and Stability*, 2013, **98**, 4, 904.

33. M. Kathalewar, N. Dhopatkar, B. Pacharane, A. Sabnis, P. Raut and V. Bhave, *Progress in Organic Coatings*, 2013, **76**, 1, 147.

34. A.S.F. Santos, J.A.M. Agnelli and S. Manrich, *Polymer Plastics Technology and Engineering*, 2010, **49**, 1–3, 254.

35. M.E. Tawfik and S.B. Eskander, *Polymer Degradation and Stability*, 2010, **95**, 2, 187.

36. R.K. Soni, S. Singh and K. Dutt, *Journal of Applied Polymer Science*, 2010, **115**, 5, 3074.

37. D.S Achilias, G.P. Tsintzou, A.K. Nikolaidis and Bikiar, *Polymer International*, 2011, **60**, 3, 500.

38. L.K. Krehula, A.P. Sirocic, M. Dukic and Z. Hrnjak-Murgic, *Journal of Elastomers and Plastics*, 2013, **45**, 5, 429.

39. A.C. Sanchez and S.R. Collinson, *European Polymer Journal*, 2011, **47**, 10, 1970.

40. Anon, *Extrusion*, 2011, **17**, 8, 24.

41. R. Shah, *International Fiber Journal*, 2013, **27**, 1, 12.

42. G. Wrobel and R. Bagsik, *Journal of Achievements in Materials and Manufacturing Engineering*, 2010, **43**, 1, 178.

43. G. Radig, *Kunststoffe International*, 2011, **101**, 2, 29.

44. SuperCleanQ EU Funded FP7 Research Project. http://www.supercleanq.eu.

45. P.D. Coates in *In-process Measurements on Polymers*, The Electrochemical Society, Penningtobn, NJ, USA. http://www3.electrochem.org

7 Testing and characterisation of recycled polyethylene terephthalate products

7.1 Introduction

The testing and analysis of any plastic product is crucial to establish whether it meets the important criteria set out in the various specifications and standards that relate to it. This is equally true for products made from virgin plastic, or those made from recycled plastic. However, in the case of recycled plastics, there tends to be additional testing requirements to ensure that certain standards have been met with regard to the sorting, separation and decontamination stages. If the intended use of the recycled plastic is for food contact applications, which is often the case for recycled polyethylene terephthalate (rPET), then an additional suite of tests must be carried out to demonstrate compliance with applicable regulations (Chapter 4).

This section is an overview of the tests and analytical methods that could be carried out on PET at all stages in its lifecycle. However, because this book is primarily about the recycling of polyethylene terephthalate (PET), the principal focus is on those that have greatest relevance for rPET. It also includes characterisation data on post-consumer PET with a view to establishing the contaminants that have built up within it during its use in service and hence must be removed by any subsequent recycling process. In doing this work, researchers have assisted the development of new recycling processes, particularly the 'super-clean' type required for food-grade rPET (Chapter 6). This work has also provided excellent data of great value used in the development of analytical methods to characterise rPET at various stages in its lifecycle, for example, the analytical method developed as a result of the Food Standards Agency (FSA) research project [1] entitled '*Develop a Post-Market Test for Recycled Food Contact Materials*' (Section 7.6.1).

Other tests that are included in this section are the standard tests required to demonstrate that a food-grade rPET product meets the regulatory requirements in the European Union (EU) (e.g., compositional and food migration tests) and has acceptable material properties [e.g., molecular weight (MW) and melt viscosity] to enable it to be processed efficiently and effectively into useful end articles (e.g., blow-moulded bottles).

7.2 General standards and characterisation tests for recycled materials and products

To ensure that the systems associated with the collection, sorting and sampling of recycled plastic are operating to a high standard, several standards have been published that cover these processes:

https://doi.org/10.1515/9783110640304-007

- ISO 15270: Plastics – Guidelines for the recovery and recycling of plastic waste.
- EN 15343: Plastics – Recycled Plastics – Plastics recycling traceability and assessment of conformity and recycled content.
- EN 15346: Characterisation of PET recyclates.
- EN 15347: Plastics – Recycled Plastics – Characterisation of plastics wastes.
- CEN/TS 16010: Plastics – Recycled Plastics – Sampling procedures for testing plastics wastes and recyclates.
- CEN/TS 16011: Plastics – Recycled Plastics – Sampling preparation.

In addition to these general and rPET standards, there are standards that address the characterisation of other recycled plastics, such as polystyrene (EN 15342), polyethylene (PE) and polypropylene (EN 15344), and polyvinyl chloride (PVC) (EN 15345).

To facilitate standardisation in this area, a standard has been written that provides guidance on the development of standards for recycled plastics: European Committee for Standardizations CEN/TR 15353 [2].

With regard to analysis of recycled plastics, as well as the samples and products that are produced from them, a large range of test methods are available to polymer analysts and test engineers. The application of these test methods enable thorough characterisation of any recycled plastic to be undertaken to establish its quality with regard to a range of requirements: physical properties, ageing resistance and the presence of contaminates. These methods include chemical analysis techniques and physical testing methods. A general overview of these techniques and their application to plastic materials and products has been the subject of several texts, including those by Forrest [3] and Brown [4]. It is not appropriate to describe these tests and techniques in detail here. Instead, a concise summary of the types of tests that workers in this area have found to be the most useful is provided.

It can be necessary to carry out some initial characterisation tests on samples of waste plastic before recycling, for example as a quality-control tool as dictated by a quality system. The chemical analysis tests carried out to obtain this type of information tend to use only relatively small amounts of samples (e.g., 10 mg to 1 g) and can be conducted on the plastic whether it is in the form of a complete product (e.g., a bottle) or flake. Two of the most common tests carried out are those to ascertain the type of generic polymer that a plastic material is composed of and tests to determine its bulk composition (i.e., types and levels of the principal ingredients). To identify the polymer(s) present in a plastic it is usual to use Fourier-Transform infrared spectroscopy (FTIR) or nuclear magnetic resonance spectroscopy. Usually, these techniques are applied to a sample that has undergone some preliminary preparation. For example, in the case of FTIR, samples can be hot pressed into a thin film, pyrolysed (Py) to remove any interference from fillers present in them, or dissolved in an appropriate solvent to cast a film.

With regard to determination of the bulk composition of a plastic sample, this is usually achieved by thermogravimetric analysis (TGA). This method is capable of providing a quantitative determination of the total amount of low-MW additives, polymer, and inorganic constituents. If further information is required on a plastic sample (e.g., type of plasticiser, inorganic filler or stabiliser system) then additional analytical work must be carried out using spectroscopic, thermal or chromatographic techniques.

To be useful, a recycling process, particularly a food-grade one, must have the ability to remove most of the contamination that the plastic has picked-up in service without reducing the important properties (e.g., MW and crystallinity) that enable the plastic to be re-made into high-quality new products such as trays and bottles. To enable a full assessment of the quality of the recycled product or material to be obtained, a large suite of tests are available and several of these are covered in Sections 7.3 and 7.4. These include general quality tests and those associated specifically with the EU regulations for recycled plastic for food use.

Several important quality-control tests can be applied to PET as it passes through the recycling process. A number of these enable polymer-related properties (e.g., MW) to be assessed, which can provide vital information on the suitability of the PET for the manufacture of specific end products (e.g., bottles). Most of these tests can be carried out on the samples in the 'as received' state and there is no requirement to produce test pieces having a specific geometry. Important examples of these quality-control tests are:

- Melt flow rate [5] ISO 1133 Plastics – Determination of melt flow rate.
- Intrinsic viscosity [6] ISO 1628-5 Measurement of the inherent viscosity of PET.
- Degree of crystallinity.
- MW distribution and MW characteristics by gel permeation chromatography* ISO 16014 Determination of molecular weight by gel permeation chromatography [7].
- Colour test (e.g., L, a and b values).
  *e.g., weight- and number-average MW

To enable the fundamental physical properties of rPET materials and products to be determined, test pieces (e.g., dumbbells) need to be prepared (e.g., by injection moulding) to allow the appropriate tests to be carried out. Some of the properties that can be assessed in this way include:

- Impact strength
- Tensile strength and elongation
- Flexural modulus
- Softening point

Standards are available for these types of physical properties, and examples include International Organization for Standardization, ISO 180 [8] for impacts tests and ISO 527-1 and ISO 527-2 [9] for tensile tests.

Finally, there are a number of other standard tests (quality and regulatory) that can be applied to the final products once they have been manufactured. For example, in the case of rPET, there is the burst pressure test for bottles. This is usually carried out by a specialised machine that meets the requirements of industry standards, such as the one published by the International Society of Beverage Technologists designated '*Voluntary Standard Test Methods for PET Bottles*'. There is also an American Society for Testing and Materials ASTM F2013-10 test [10] that can be used to quantify the amount of residual acetaldehyde (AC) in PET bottles and performs.

The results obtained by these types of tests will enable a final decision to be made as to whether the rPET is capable of being used for the intended application.

7.3 Tests to characterise and assess the quality of recycled polyethylene terephthalate

Section 7.2 provided a brief overview of the range of tests that can be applied to rPET and other recycled plastics. This section contains published examples of tests that have been applied to rPET for specific purposes [e.g., its detection in mixtures with virgin polyethylene terephthalate (vPET)]. To complete the picture, the tests required if the rPET is to be used for food contact applications are covered in Section 7.4.

One of the perennial questions posed to the polymer analyst is: is it possible to distinguish between virgin plastic and recycled plastic? Although appearing simple at face value, it is extremely difficult question to provide a definitive answer to, due to several factors, irrespective of the type of plastic in question. These factors include the limited number of available diagnostic substances; the potentially infinite number of virgin plastic/recycled plastic blend ratios; and the influence that different recycling conditions and processes have on the final properties of the recycled plastic. However, as Romao and co-workers [11] pointed out, it is important for industry to have an experimental method to detect rPET in a batch of product. They evaluated the possibility of using differential scanning calorimetry (DSC) for this purpose. They studied the influence of the presence of post-consumer bottle-grade PET and thermomechanical processing on the thermal properties of bottle-grade PET. This was possible because when virgin resin is subjected to thermomechanical processing (e.g., production of pre-forms, production of soft drink bottles, recycling or preparation of rPET/PET blends), the crystallisation rate is affected and a clearly defined crystallisation peak (T_c) is observed.

This Brazilian group found that bimodal melting temperature (T_m) behaviour was observed for samples subjected to processing and that if rPET was present in the material, then T_m bimodal behaviour was followed by a narrowing and a shift of the T_c to higher temperatures. Therefore, they concluded that the crystallisation rate, the T_m and the T_c are the principal thermal properties that can be used to distinguish between vPET and rPET that have been subjected to thermomechanical processing.

Another technique that has been evaluated to see if it can be used to detect rPET is matrix-assisted laser desorption ionisation–mass spectrometry (MALDI–MS). Romao and co-workers [12] showed that this technique can provide valuable information on the thermomechanical degradation of bottle-grade PET. They used MALDI–MS to monitor an admixture of post-consumer bottle-grade rPET and vPET, and the effects of thermomechanical degradation on the chemical properties of vPET. Principal component analysis of MALDI–MS data was used to classify the samples into four groups with specific features:

a)   rPET with intrinsic viscosities of 0.80 or 0.65–0.60 dLg^{-1};
b)   Processed or vPET with the same intrinsic viscosity;
c)   vPET from vPET which contains rPET; and
d)   vPET from different manufacturers.

They concluded that the data obtained by MALDI–MS was capable of being used as a quality-control tool for bottle-grade PET resins and could, therefore, be used to prevent fraud and the illegal use of recycled bottle-grade PET.

In addition to determining the level of contaminants in a recycling stream or a recycled product, or characterising the physical properties of the final rPET material, analytical tools can also be used to selectively quantify residual additives (e.g., stabilisers) in recovered PET before it passes through a recycling process. This can help processors to decide whether or not important additives have been depleted to such an extent during the products initial life that additional levels need to be added to ensure that it will perform satisfactorily during the recycling stages and throughout its second life [13].

The systems used to remove contamination from post-consumer PET can have an adverse effect on its properties due to processes such as hydrolysis. Analytical work can be carried out to determine the extent and impact of these adverse effects. Ptieek and co-workers [14] studied the effect of the alkali pretreatment of post-consumer PET on the properties of PET flakes. To achieve this goal, PET flakes were washed at two temperatures (70 and 75 °C) and at various time intervals (15, 18, 21, 25 and 30 min). At the end of treatments, all samples were characterised by FTIR, DSC and contact-angle measurements. Results showed that, during alkali treatment, partial depolymerisation of the PET took place, which resulted in the formation of various types of oligomers with hydroxyl and carboxyl end groups. A decrease in the intensity of three characteristic IR vibrational bands (CO at wavenumber 1,717, COO at 1,265, and CH_2 at 722 cm^{-1}) with extended contact time was observed. The formation of hydroxyl groups (wavenumber, 3,428 cm^{-1}) was also observed as the degree of depolymerisation increased due to the use of the longer contact times and the higher contact temperature. DSC work revealed multiple melting peaks in some of the samples which the authors thought could be due to partial melting and re-crystallisation, or to new polymer fractions of low MW. The contact-angle measurement showed that the treated PET flakes had a lower contact angle than the unexposed control samples as a result of the changes in the chemistry at the surface. The increase in hydrophilicity was thought to be due to an increase in the number of hydroxyl and acid groups present. Overall, the team concluded that a partial depolymerisation of the PET flakes occurred during alkali treatment, but that the material retained its good properties and was still suitable for recycling.

It is also possible to carry out continuous assessment of a process line (e.g., an extrusion line) to ensure consistent performance and, hence, quality. Such an operation is referred to as 'online monitoring'. In an article in the *International Fiber Journal*, Gneuss [15] of Gneuss Incorporated described how the company's melt-pressure transducers can be used to achieve continuous online measurement and control of processing parameters in PET recycling operations. The article also described the TF range of T_m sensors, and a technical layout of an online viscometer was provided.

The new technologies that are now available are opening up novel ways in which specific properties of rPET products can be determined. For example, an article in *PETplanet Insider* [16] described a free app for iPhone/iPad/Android platforms that was designed to calculate the intrinsic viscosity of PET resin blends, including those that featured rPET. The calculation process was described as going through two stages. The first screen requires the user to enter in the volume percentage and intrinsic viscosity for the rPET and vPET resin being processed. The new intrinsic viscosity and MW of the resultant blend before processing is then calculated. The user is then directed to a second screen, which features a PET hydrolytic degradation nomograph, to input moisture information, in ppm, and then the intrinsic viscosity that the blend will be after processing is calculated.

Computer-based assessments can also be carried out using commercially available models. For example, Prabhu and co-workers [17] presented a paper that described the results of an investigation carried out using Taguchi methodology into the effects of the processing parameters on the properties of rPET/fly ash composites. The group generated composites of rPET and low-cost fly ash cenospheres (FAC) and then used compression moulding to produce samples that could be tested for their wear resistance and flexural modulus. The possible end use for these composites was in the manufacture of plastic gears in low-cost applications such as children's toys. Taguchi methodology was employed for the experimental design and the analysis of the data. Results indicated that the proportion of the FAC in the samples, the moulding pressure and the mould temperature were the major contributors to the eventual strength and wear performance of the composites. The use of response surface methodology revealed that ≈5.4–7% FAC content, together with a moulding pressure of 11.6–14.4 MPa, were the optimum parameters for flexural strength and wear rate.

7.4 Tests on recycled polyethylene terephthalate material and products intended for food contact applications

7.4.1 Decontamination of post-consumer polyethylene terephthalate by the recycling process

For the recycling of plastics into food contact materials and articles, EU Regulation EC 282/2008 states that the efficiency of the recycling process to remove contamination must be assessed using the European Food Safety Authority (EFSA) 'challenge test'. The details of this test, which involves 'spiking' post-consumer PET that is to be recycled with marker compounds that are representative of the types of organic compounds (i.e., with regard to polarity and MW) that the PET articles could come into contact with during first use (under normal conditions and due to abuse), have been published in the European Commission report *'Guidance and Criteria for Safe Recycling of Post-Consumer PET into New Food Packaging Applications'* [18]. A recycling process must be capable of reducing the concentration of these marker compounds to a specified minimum level for the process to meet EFSA requirements and Regulation EC 282/2008 (Sections 4.2.4.2 and 4.2.6).

For this evaluation process EFSA has developed a conservative concept to protect consumers, which is based partly on mathematical calculation using a model that overestimates migration by a factor of 5. This factor applies to molecules with a small MW, such as toluene, but for higher-MW substances (e.g., benzophenone) the overestimation factor will be even higher. The reason for this overestimation is that the currently used migration model is based on a fixed activation energy of diffusion. Conversely, the curve of the maximum bottle wall concentration, calculated using the current migration model, increases much too gradually with the MW. New developments that have taken place in migration modelling consider more precisely the activation energies of diffusion. Hence, using the more realistic diffusion coefficients has a significant influence on the EFSA evaluation criteria [19]. It was due to these concerns that the research project to develop an analytical method for some of the most common organic contaminants in food-grade PET was commissioned by the FSA in the UK (Section 7.6.1).

As mentioned in Section 7.6.2.1, *'Scientific Opinion on the Assessment of Food-Grade PET Recycling Processes'* was published by the EFSA Panel of food contact materials, enzymes, flavourings and processing aids in *EFSA Journal* in 2011 [20]. This document was entitled *'Scientific Opinion on the Criteria to be used for Safety Evaluation of a Mechanical Recycling Process to Produce rPET Intended to be used for Manufacture of Materials and Articles in Contact with Food'*.

7.4.2 Existing quality standards and test methods for food contact materials

To assist industry to demonstrate that food contact materials and articles meet the requirements of EU food contact regulations, such as the Plastics Regulation EU 10/2011 (Chapter 4), a series of European standards have been published by CEN [21].

With respect to regulatory tests (i.e., those that are mandatory because they are required to pass EU food contact regulations) one prime example for PET (in vPET and rPET forms) is determination of AC. This compound is of interest because its presence in PET bottles can affect the taste of the drink that is packaged within them if it migrates. One test that is often used to quantify AC is the ASTM method F2013-10 (Section 7.2). AC is also targeted because its potential toxicity means that it has a specific restriction of migration limit in the Plastics Regulation EU 10/2011 of 6 mg/kg (of food or food simulant). The sample of food simulant (or food sample) for this test will usually be prepared using the final PET product (e.g., bottle) and taking into account representative contact times and temperatures for its end use.

In addition to specific substance tests, food-grade PET products must also pass the general EU overall migration limit for plastics of 10 mg/dm^2 (or 60 mg/kg of food) as described in EU 10/2011. The products must also fully satisfy the requirements of Article 3 of the EU Framework Regulation EC 1935/2004, which involves organoleptic testing for determining if the PET product imparts a taint, odour or colour change to food (Section 4.2.3.1).

7.5 Contaminants and potential migrants in recycled polyethylene terephthalate

Several studies have been carried out to identify and, in some cases, quantify, the wide range of contaminants that may be present in recovered post-consumer PET and rPET. These contaminants can be of interest due to the impact that they have on performance criteria, such as ageing or, if the rPET is going to be used in food-grade products, because they could potentially migrate into food. This section provides an overview of the work that has been published in this area, some of which has influenced regulations that have come into force (e.g., EU Recycling Regulation EC 282/2008 for recycled plastics) (Section 4.2.4.2).

7.5.1 Heavy metals

As described in Chapter 4 there are regulations in place in several countries and regions of the world (e.g., the EU) to ensure that rPET for food contact applications meets a certain standard in terms of purity. However, there is still a need for constant vigilance to ensure that effective sorting, separation and recycling processes are being employed, and this has been a theme recurring several times in this book. This requirement has been illustrated by studies carried out to determine the purity of rPET products, particularly food-grade rPET products. An example of such an investigation was the one carried out by researchers at California Polytechnic State University in the US [22] to survey the heavy-metal contamination present in rPET for food packaging. The group found heavy metals in the packaging, which they believed was due to deficiencies in sorting and recycling processes. They suspected that the source of the contamination was the increased use of recycled plastic from international suppliers and its contamination with electronic waste. The technique of inductively coupled plasma–atomic emission spectroscopy was used in conjunction with the ASTM E1613-04 test method to quantify nickel, chromium, cadmium, antimony and lead in 200 samples of extruded sheet and thermoformed products. Of the 200 samples, 29 samples were found to be contaminated with heavy metals. Chromium and cadmium were found in all 29 samples; nickel, lead and antimony were found in >90% of the sample replicates. With regard to the amounts present, nickel was present at an average concentration of 11.59 ppm and lead at 0.15 ppm. It was noted by the group, however, that the total contamination in all 29 samples was well below the threshold level set for the incidental presence of heavy metals in packaging materials by the Californian Toxics in Packaging Prevention Act of 2006. They did not carry out migration work to determine the level of this contamination for potential migration into food.

Other workers have targeted the antimony that is commonly used as a catalyst in the polymerisation process employed to produce PET. PET is used extensively for food-grade bottles (Chapter 8) and there is potential for the residual antimony to migrate into the food products (e.g., juices, soft drinks or water) placed into these bottles. In Brazil, Shimamoto and co-workers [23] reported that there are concerns expressed in the literature regarding the toxicity of antimony. Hence, they carried out work to determine its level in PET bottles. The team used a fast and direct method based on X-ray diffraction (XRF) to quantify the antimony, and also sulfur, iron and copper, in the bottles. Results showed that antimony was present at 2.4–11.0 mg/kg in the 20 samples that were analysed. In addition, the team showed that, by coupling the multiple-element technique to chemometric treatment, it was possible to use the iron content to differentiate between bottle-grade PET and rPET blends. Another group of Brazilian researchers [24] used a combination of XRF and chemometric analysis to determine metals in PET samples. Using these substances as a diagnostic tool, these workers demonstrated, by the analysis of vPET, rPET and their blends, that it was possible to determine the presence of rPET in bottle-grade resins. They also showed that it was possible to differentiate between different methods of recycling, such as mechanical processes as opposed to chemical processes, such as esterification or transesterification. They used the data obtained to construct models that they suggested were able to predict the weight percentage of rPET in bottle-grade PET.

7.5.2 Substances formed due to the decomposition of polyethylene terephthalate

AC is one of the most common thermal degradation products of PET. It can be found in vPET and rPET, but it is usually at a higher concentration in rPET because the products have been in service and, therefore, subjected to degradation agencies such as ultraviolet (UV) light, and the bottles have been shredded, which will have resulted in some additional 'heat history'. Franz and co-workers [25] determined levels of AC in rPET to be 18.6–86.0 mg/kg and found that, once these recycled materials were re-extruded, the level could be reduced to 1–20 mg/kg. Two other compounds that originate from the PET polymer and which can be detected in vPET and rPET are 2-methyl- 1,3-dioxolane (MD) [a condensation product of AC and ethylene glycol (EG)], and EG (a residual monomer) [26].

A Japanese group [27] described an analytical technique for rapid determination of terephthalic acid (TPA) in the hydrothermal decomposition products of PET by thermochemolysis–gas chromatography (GC) in the presence of tetramethylammonium acetate. They stated that the technique can be applied directly to the samples to determine the decomposition products obtained from the hydrothermal recycling of PET and, on the GC chromatograms obtained, a sharp peak of the TPA will be observed clearly. By using the size of these peaks, it was possible to precisely and rapidly determine the amount of TPA present in samples without time-consuming sample pretreatment steps.

Badia and co-workers [28] carried out an investigation into the role of crystalline, mobile amorphous fraction (MAF) and rigid amorphous fractions (RAF) in the processing performance of rPET. To simulate the thermomechanical degradation that results from the mechanical recycling of PET, the group exposed the material to successive injection-moulding cycles. They found that degradation reactions during these processing steps provoked chain scissions and a reduction in MW, and they believed that these were driven mainly by the reduction of diethylene glycol to EG in the flexible domain of the PET backbone, and the formation of hydroxyl terminated species with a shorter chain length. Work was carried out to quantify the microstructural changes that occurred as a result of these reactions taking into account a three-fraction model involving crystalline, MAF and RAF. Results showed that a large increase in the RAF occurred to the detriment of the MAF and that the crystalline fraction remained nearly constant. The group also carried out an analysis of the melting behaviour, the segmental dynamics around the glass–rubber relaxation, and the macroscopic mechanical performance. Also, they described the role of each of the three fractions in the rPET using thermal, viscoelastic and mechanical properties. Data revealed that these important properties had been affected adversely by thermomechanical degradation even after the first injection-moulding cycle.

7.5.3 Substances absorbed by polyethylene terephthalate during use

7.5.3.1 Flavour compounds and other species from foodstuffs

A very common flavour compound detected in rPET is limonene (Lim) and it has been found in the material in the range 0.1 to 20 mg/kg by Franz and co-workers [25, 29]. In addition to Lim, p-cymene was also detected in many samples by Nerin and co-workers [30]. This work was carried out qualitatively, but it was estimated that the concentration of p-cymene was around 5-fold lower than that for Lim. Nerin also reported finding aliphatic aldehydes, such as 2-decenal and 1-octadecenal, in bottles of flavoured soft drink and Bayer [31] reported finding flavour-like terpenoid compounds [e.g., β-burbonene and (Z) β-farnesene].

The differences in the extraction profiles that can be obtained from PET bottles used to store different food products was illustrated by Mancini and co-workers [32]. They compared post-consumer PET bottles that had been used to store oil and a soft drink. The PET container that had been used to store oil had a much more complicated profile of species, with 35 organic species being detected (5 of which were non-volatile) and 7 metals.

7.5.3.2 Substances present as a result of the misuse of polyethylene terephthalate products

Another source of contaminants in post-consumer PET arises due to the misuse of PET bottles (e.g., for storing engine oil and household chemicals). Examples of the types of chemicals that can be detected in rPET as a result of the storage of household chemicals in the products before recycling include solvents. These types of contaminants are intermittent, as are those described in Section 7.5.4. Franz [25] reported that they are present mostly at low concentrations (1.4–2.7 mg/kg), though a substance detected in this work thought to be toluene (from its retention time) was present in one flake sample at 450 mg/kg. In the same study, xylene isomers were also detected in a few samples of flake at concentrations of 50–200 mg/kg.

Two estimates of the frequency of the misuse of PET bottles are from Bayer and co-workers [33]. They reported that the frequency of misuse of PET bottles is one misused bottle per 10,000 uncontaminated bottles, and Franz [25] estimated the figure to be one bottle per 3,000.

7.5.4 Substances picked up by the polyethylene terephthalate during recycling operations

It is possible during the collection and sorting of PET bottles for other plastics to remain in the waste stream and these will then contaminant the PET flakes exiting the shredder. Most plastics will contain some additives introduced into the material for technological reasons and some (particularly PVC) contain a considerable number, and these can appear in the rPET as contaminants.

Two common examples of these additives are:

a) Plasticisers such as phthalates (dibutyl phthalate, dioctyl phthalate, butyl benzyl phthalate) and adipates (e.g., diethyl hexyl adipate).

b) Slip additives such as erucamide and oleamide (used extensively in polyolefin films).

Due to the random nature of their inclusion, their appearance in rPET is sporadic and they tend to be present at very low concentrations. Franz [25] suggested levels around the limits of detection of analytical techniques (e.g., 0.05–0.2 mg/kg) and Nerin reported an example of the plasticiser dioctyl adipate being present at 0.5 mg/kg [30].

7.5.5 General studies

Franz and co-workers [25] published the results of work aimed at establishing a statistical overview of the nature and extent of contaminants in post-consumer PET. This information resulted from the European project FAIR-CT98-4318 'Recyclability' [34]. An important general finding concerning migrants in PET, and their potential to migrate, was commented on by Franz and co-workers [25]. They suggested that, owing to the extreme low diffusively of PET, substances with a MW >350 g/mol are virtually immobilised in the PET matrix and so will migrate very slowly.

The Fraunhofer Institute IVV was the task leader for the PET section (Section I) of this FAIR-CT98-4318 project [34]. Workers at this institute developed analytical tests for flake, pellet and packaging products to detect the level of decontamination of the PET material and the migration of any species from PET packaging into food products. For the analysis of the PET itself, they used headspace GC, with the samples being heated and the volatile products being detected by their characteristic retention times. The substances listed below were typically identified in post-consumer PET:

1. AC
2. MD
3. EG
4. Lim

The first three of these compounds are found in all PET resins and originate from the polymer itself; the last substance is a common compound used to flavour soft drinks.

The results of the analytical work showed that after decontamination using a 'super-clean' process (Chapter 6), all of the contaminant species had been removed from the rPET flake and, in fact, it contained even fewer potential migrants that conventional vPET. The project report also described the migration modelling of substances based upon MW and showed that the inverse relationship between the amount that migrated after 10 days at 40 °C and MW reached a limit beyond 350 g/mol. This finding was because beyond this MW value the substances were virtually immobilised in the PET matrix.

Workers such as Nerin [30] and Bayer [31] reported the results of screening studies on several rPET samples and a wide range of chemical compounds detected by sensitive analytical methods such as headspace GC–MS. For example, in addition to the classes of compound already covered in the sections above, the following were found:
a) Aromatic aldehydes
b) Esters
c) Aliphatic acids
d) Alkanes
e) Ketones
f) Ethers

These studies tend to be carried out in only a qualitative manner due to the large number of samples analysed and so concentration values (e.g., in ppm) are not reported.

7.5.6 Migration studies

As discussed in Chapter 4, in this section and elsewhere in this book, challenge tests to test the efficiency of recycling processes for food contact plastics are mentioned in the EU Plastics Recycling Regulation EC 282/2008. Several published papers focused on migration studies undertaken using samples of PET spiked with surrogate chemical compounds, such as the one published by Widen and co-workers [35] in which the following substances were used:
a) Lim
b) Benzaldehyde
c) Benzophenone
d) Anethole

The use of 'cocktails' of surrogates to determine the decontamination efficiency of new or modified recycling processes were discussed by Welle [26] and in the *TNO V6633 Report* published in a 2005 TNO Report [36].

Several researchers have looked into the migration of organic compounds from vPET and rPET into various contact media (e.g., Monteiro and co-workers [37] and Hinrichs and Piringer [38]). They found that the diffusion laws governing migration were not affected by the recycling process.

One example of the information in the literature on the migration of oxidation species from PET is the paper by Mutsuga and Kawamura [39] on the migration of formaldehyde and AC into mineral water from PET bottles.

The FSA in the UK has provided funding for many research projects concerned with food contact materials over the last 20 years. The results that were obtained from one of the more recent projects are described in Section 7.6.1. Another example of an FSA project relevant to PET is FSA project A03049. This was project was entitled 'An investigation of functional barriers currently used by the food industry and an assessment of their efficacy'. The final report for this project, which looked at the performance of functional barriers in food packaging, was published in March 2009 [40]. Several migration scenarios were studied during the lifetime of the project and, one that had relevance for PET, was the use of a PET trimer (MW: 576 g/mol) as a surrogate compound. PET was immersed into a 'spiking' solution containing this trimer. Then, this layer was placed against the PET film on the non-food contact side of the functional barrier and migration experiments carried out. In this set up, to get into the food simulant, the PET trimer had to migrate out of the spiked layer, through the non-food contact side PET, through the functional barrier, and finally through the PET film on the food contact side. Experiments were also carried out without any functional barrier. Sunflower oil (simulant of a fatty food) and aqueous food stimulants were used, and the PET trimer determined by high-performance liquid chromatography with a UV detector. Results showed that very little migration of the PET trimer. The only positive result was obtained when the spiked layer was in direct contact with sunflower oil at a high temperature (100 °C). Migration did not occur if a functional barrier was present.

Waste and Resources Action Programme (WRAP) is another UK organisation that has funded many research projects, as demonstrated by the number of times they are mentioned throughout this book. One project concentrated on the use of rPET for food contact products and was carried out between August 2004 and February 2006 [41]. The recycling of the post-consumer PET that was used in the project was carried out by the Closed Loop Recycling company in London, and the rPET generated was incorporated into products that were marketed by the retailers Boots and Marks & Spencer. The project sought to demonstrate the viability of using rPET in retail packaging and was regarded by WRAP as being successful. This was because, before the project, no major retailer in the UK produced PET packaging containing a significant level of recycled material derived from post-consumer recyclate. Two principal types of food packaging were produced from the rPET during the project:

- Thermoformed products (e.g., salad bowls and lids)
- Injection blow-moulded bottles (e.g., for containing juice)

As part of the due diligence process, migration testing and headspace measurements were carried out on some of the rPET products at the Fraunhofer Institute for

Process Engineering and Packaging IVV. Migration testing was carried out on thermoformed salad bowl products using 95% ethanol as a 'worst case' food simulant, and the presence of volatile substances was detected by headspace GC. Products made from vPET were used as controls. The headspace results for the virgin and recycled products were similar, but the chromatograms for the rPET products contained additional peaks at longer retention times where flavour molecules such as Lim and cineole are usually found. All of the peaks in this region were <1 ppm. With regard to migration tests, the results with thermoformed salad bowls containing 50% of rPET were 16-fold lower than the maximum overall migration limit in the EU for plastics of 10 mg/dm^2. Overall, the analytical results showed that materials containing 50% of rPET meet the standards of performance expected for food contact packaging adequately.

In addition to the analytical work described above, examples of the rPET salad bowls and lids were delivered to Geest for filling with salads and shelf-life stability tests. The products passed the storage and organoleptic tests though there was some condensation on the lids, which was due to the absence of the anti-misting treatment that is usually applied to this type of product. Following these successful project results, Marks & Spencer gave permission for the launch of the thermoformed packaging containing 50% of food-grade rPET.

With respect to the rPET juice bottles, headspace evaluations were not carried out on these products, but bottles containing rPET were submitted to Fraunhofer for migration testing. Again, 95% ethanol was used as the food simulant and the overall migration was 0.1 mg/dm^2 (100-fold lower than the EU limit of 10 mg/dm^2). Instead of organoleptic testing, Orchard House Foods Limited decided to measure the microbiological influence of the rPET in the juice bottles. Results showed that all the bottles tested were free of contamination by yeasts and moulds. These results enabled the introduction of food-grade rPET into bottles at levels of ≥30% for juice bottles.

7.5.7 Influence of contaminants on ageing of polyethylene terephthalate

A group of Croatian workers [42] examined the impact of contaminants in post-consumer PET bottles on the decomposition of polymer once it had been recycled. To achieve this goal, in addition to analysing samples of vPET and rPET, samples were also analysed that had been contaminated deliberately with substances known to degrade PET. Several approaches were used by the group to determine the thermal degradation products of the PET samples: Py–GC–MS, temperature-programmed evolved gas analysis–MS, and TGA. The chemical and structural information obtained from the work suggested that, if contaminants were present in the PET samples, there was some change in their mechanism of decomposition.

7.5.8 Summary of the contaminants detected in polyethylene terephthalate

As shown above in Sections 7.5.1–7.5.7 a range of contaminants have been detected in rPET from the work that has been carried out. These studies have shown that the major species present in post-consumer PET originate from two sources:
1. The PET polymer; and
2. Flavouring substances present in the food that has been packaged in the PET.

Some workers have detected a large range of other compounds, but these have all been present at very low levels and some have originated from small quantities of contaminated PET (due to misuse of the containers), or small amounts of other polymer-based waste entering the PET waste stream. The low level and intermittent and sporadic nature of these compounds makes them unsuitable for use as marker compounds in rPET studies.

A summary of the potential migrants reported in post-consumer PET is shown in Table 7.1.

Table 7.1: Potential migrants in post-consumer PET.

Potential migrant	Range found in post-consumer PET
AC	18.6 to 86.0 mg/kg
EG	N/A
MD	N/A
Lim	0.1 to 20 mg/kg
p-Cymene	≈0.02 to 4 mg/kg
IT	≈0.02 to 4 mg/kg
Numerous organic compounds	≈0.05 to 0.5 mg/kg

N/A: Cited by Franz and co-workers [25] as detected but a concentration not provided
IT: 4-Isopropyltoluene

It is tables such as these that are often used as the starting point for the development of new analytical methods to identify and quantify contaminants (i.e., marker compounds) in rPET. Two examples of such methods, both developed as a result of the activities of two recent research projects, are described in Section 7.6.1 and 7.6.2.

The term 'marker compound' is used widely, and one definition for it is provided below.

Marker compounds for recycled food contact materials such as PET are relatively low-MW compounds which are found consistently in the material, and can migrate into food when it is used for the packaging of food products. They vary in

several ways, for example in their potential toxicity and whether they are single compounds or a mixture of isomeric forms of a compound. Another very important way in which they vary is in their origin, which is dependent on:
a) The material itself and its chemical properties and resistance to influences such as thermal degradation.
b) The technology associated with the manufacture of the food contact packaging from a particular material.
c) The types of food products that they contact during their original use and the conditions under which they contact these food products.

The term 'marker compound' can also be used for compounds which reflect, in MW and polarity, the types of substances that can be found in a post-consumer plastic and which can also be added deliberately to the material to test the decontamination efficiency of a recycling process designed to produce food-grade recyclate. These compounds are often referred to as 'surrogate' or 'challenge compounds' and are used in the challenge tests required by the EFSA and the US Food and Drug Administration (Chapters 4, 6 and Section 7.6.2).

7.6 Development of new analytical methods for determining contaminants in recycled polyethylene terephthalate

7.6.1 FSA research project FS241007 analytical method

This project that was funded by the FSA, had the title '*Develop a Post-Maker Test for Recycled Food Contact Materials*' and ran from 2011 until 2013. The final report, dated 11th March 2014, was published by the FSA on their website [1].

In this FSA project, five contaminants (i.e., marker compounds) that could be present in rPET were selected for determination:
1. AC
2. MD
3. EG [also called monoethylene glycol (MEG)]
4. Lim
5. IT

To prepare the PET samples (all types, see below) for analysis, initially they were grounded cryogenically through a 2-mm ring sieve at 16,000 rpm. Due to the volatility of the marker compounds, headspace GC–MS was the preferred choice of analytical technique. However, the initial development work showed that it posed problems for quantification of the MEG. Hence, a method based on headspace GC–MS was developed for the AC, MD, Lim and IT, and a method based on accelerated solvent extraction (ASE) using acetone, followed by direct injection GC–flame ionisation detector (FID), was developed for the MEG. Both of the methods were validated for their respective marker compounds to ensure that accurate, reproducible data could be obtained.

An example of a headspace GC–MS chromatogram obtained on a rPET sample is shown in Figure 7.1. The peaks due to the four marker compounds have been assigned.

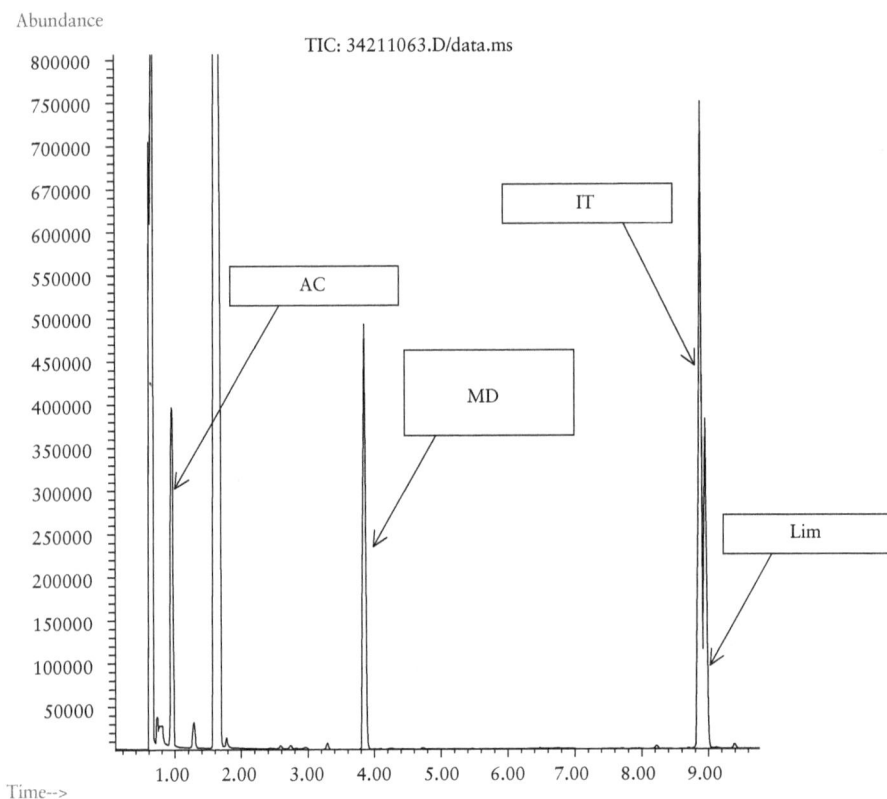

Figure 7.1: Example of the headspace GC–MS chromatogram obtained for a rPET sample. (The large peak at a retention time of ≈1.8 min is acetone and the peak at ≈0.7 min is for air).

An example of a GC chromatogram obtained on an ASE extract of a rPET sample is shown in Figure 7.2. The peak due to MEG has been assigned.

The rPET samples for this analytical work were obtained from UK-based PET recycling companies and were representative of the following stages in the recycling process:
1. Post-consumer PET bottles that had gone through the following:
   – Hot caustic pre-wash to remove labels and provide a general clean up.
   – Grinding to flake.

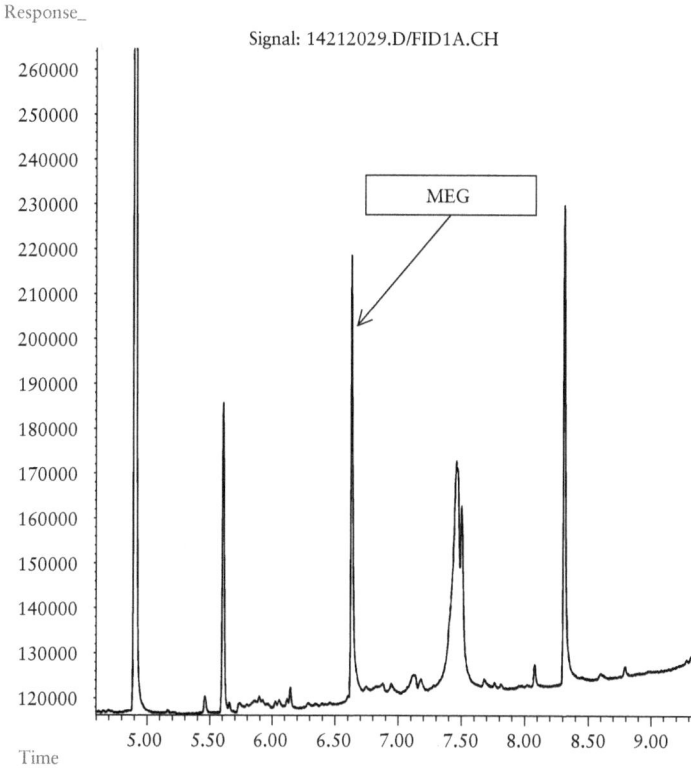

Figure 7.2: Example of the direct injection GC–FID chromatogram obtained on a rPET sample.

2. Flake that had undergone a caustic pre-wash and then been treated with a hot wash/chemical clean-up process to remove contamination [high-quality washed flake (HQWF) for use in non-food products].
3. Pellets of food-grade rPET produced from material that had been through a 'super-clean' process.
4. PET bottle preforms produced from the food-grade rPET pellets blended with 50% vPET.
5. Food-grade PET bottles 'blow-moulded' from the food-grade (rPET 50:50 vPET) bottle preforms.

The duplicate results obtained from the samples described above are shown in Table 7.2.

An additional set of samples from one of the recyclers enabled evaluation of the variation in the concentration of the marker compounds that occurred over a 5-day period. The results obtained for these samples are summarised in Table 7.3.

Table 7.2: Results obtained on the five marker compounds present in PET samples taken from different stages in the recycling process.

Sample type	Concentration (µg/g)				
	AC	MD	IT	Lim	MEG
Flake after caustic pre-wash	8.21 and 4.47	4.07 and 3.53	<LOQ	0.85 and 1.46	19.00 and 18.11
Flake after hot chemical wash	4.05 and 3.38	3.60 and 4.23	<LOQ	0.79 and 1.20	17.57 and 17.65
Granule from 'super-clean' process	5.34 and 6.23	0.81 and 0.93	<LOQ	<LOQ	18.87 and 20.91
Preforms from 'super-clean' granules	6.15 and 6.18	1.24 and 1.10	<LOQ	<LOQ	21.94 and 24.32
Bottles from preforms	2.80 and 2.05	1.83 and 1.54	<LOQ	<LOQ	23.97 and 23.69

LOQ: Limit of quantification

Table 7.3: Results obtained on the five marker compounds present in PET samples collected over a 5-day period.

Sample type	Concentration range (µg/g)				
	AC	MD	IT	Lim	MEG
Flake after caustic pre-wash	5.33–7.46	3.56–4.33	<LOQ–0.13	0.78–2.22	19.40–28.68
Flake after hot chemical wash	3.61–4.91	3.60–4.56	<LOQ–0.12	0.79–1.47	17.06–25.69
Granule from 'super-clean' process	4.16–7.46	0.76–1.02	All results <LOQ	All results <LOQ	13.88–21.42

The duration of the project (2 years) also enabled assessment of the variation of these five marker compounds in the rPET processed by a recycler over a 7–8-month period. Results showed a reasonable level of consistency in the levels at which most of them were observed. One of the exceptions was the flavouring compound Lim, which was found to vary to a greater extent, presumably due to the level present in the packaged product that a high proportion of the PET containers had been in contact with.

During the project a targeted food-migration study was done on final, food-grade PET products to determine how much of one of these marker compounds, AC, migrated into food simulants under representative conditions. This marker compound was selected due to its toxicity and because it was present at a reasonably high level in the products (Tables 7.2) and so able to migrate to a detectable level into the simulant.

One of the recyclers provided several food-grade PET bottles manufactured using 75% of their food-grade rPET for this stage of the project. Two other types of recycled food-grade products were also provided: circular 'salad pots' and pyramidal 'deli trays'.

Initially, the level of AC in these products was determined using the analytical method that had been developed, and the results were:

– Bottles, 4.89 µg/g
– Circular pots, 4.84 µg/g
– Pyramidal pots, 5.67 µg/g

For the food-migration work, the PET bottles were tested as follows:
1. Five bottles filled to the neck with 3% acetic acid for 10 days at 60 °C.
2. Five bottles filled to the neck with 20% ethanol/80% distilled water for 10 days at 60 °C.

For the PET deli trays, a 1-dm^2 portion was removed from each tray (circular and pyramidal type) and mounted onto a stainless-steel cruciform support in a glass test tube with 100 ml of the food simulant (see below) that had been pre-warmed. Then, the test tubes were placed into an oven set at the correct temperature for the appropriate time (see below). The following simulants and immersion conditions were used:
1. Olive oil for 10 days at 40 °C
2. 3% Acetic acid for 10 days at 40 °C

Then, the amount of AC present in all of the food-simulant samples was determined by headspace GC. Results showed that the levels of AC that had migrated from the bottle and the two types of deli tray samples were well below the specific migration limit (SML) (6.0 mg/kg) stated in the Plastics Regulation EU 10/2011.

The overall conclusions for the FSA project were:

– Variation in the level of marker compounds through a recycling process: The results showed some trends for rPET. The markers associated with use in service (e.g., flavouring compounds) were found to decrease, whereas those associated with the material itself (e.g., AC) showed some variation depending upon the stage in the recycling process.
– Variation in the level of marker compounds in the rPET over time: The data obtained for rPET samples processed over a 5-day period within 1 week, and over a 7–8-month period showed that, for most of the marker compounds, there was a measure of consistency, although for some (e.g., Lim) greater variability was noticeable.
– Food-migration data for AC: A limited amount of food-migration testing using rPET products was carried out. Results obtained with food simulants under representative conditions showed that migration of AC occurred but was below the SML for the compound in the Plastics Regulation EU 10/2011.

7.6.2 SuperCleanQ EU FP7 project analytical methods

7.6.2.1 Method for determining contaminants in recycled polyethylene terephthalate materials and products

The SuperCleanQ project [43] ran from 2011 until 2014 and had several objectives, all of which concerned the recycling of PET for food contact products. One of the objectives was to develop an analytical method for marker compounds in rPET materials and products (which is covered in this section). The other objective was to develop an inline monitoring method for detecting contaminants in rPET processing lines (Section 7.6.2.2).

The headspace GC–MS analytical method developed during the SuperCleanQ research project has similarities to one of the methods developed during the UK FSA project (Section 7.6.1). However, whereas the FSA approach was designed to be used as a post-market tool, and the two separate methods were not published as a British or European standard, one of the prime objectives of the SuperCleanQ project was to develop and validate a versatile, single analytical method that could be published as a CEN Technical Specification for use by the PET recycling industry in Europe. The project was successful in this goal and the SuperCleanQ method was published as a CEN Technical Specification in 2015 [44], giving it a measure of authority and acceptance. The method, which was validated for the determination of six contaminants (i.e., marker compounds), was chosen because the marker compounds are the most commonly found in PET and/or of regulatory interest. These substances are:

a) AC
b) Ethyl acetate
c) Ethanol
d) MD
e) Hexanal
f) Lim

The method developed for the analysis of these compounds is based on headspace GC-MS, which is readily available to industry as an in-house facility or through an analytical service provider. It is capable of identifying and quantifying the six compounds in one operation on PET samples that are in any physical form (i.e., flake, granule, bottle, tray). It is, therefore, a cost-effective quality-control tool that is flexible enough to be used on PET that has been sourced from any of the principal stages present in a PET recycling process. For example:

a) The collection of post-consumer PET plastic products (e.g., bottles) from road sides, and bottle banks.
b) The initial cleaning of the PET flake to remove residues from the labels and label adhesive (e.g., using a 'caustic wash' process).
c) 'Deep cleaning' using mixtures of chemicals to create HQWF.

d) Conversion of HQWF into pellets of food-grade rPET by a super-clean process.
e) Conversion of the food-grade rPET pellets into food-grade rPET products, such as bottles and trays.

The principal limitation of the method is that because it has been targeted to identify and quantify the six contaminants listed above it is not able to assess contamination in the rPET that could arise from other sources:

1. Other breakdown products, reaction products or chemical contaminants originating from the PET polymer itself or components used in its manufacturing process (e.g., the polymerisation stage).
2. Contamination originating from sources such as other polymers/ labels/adhesives/inks and printing or decoration on the outside of the PET packaging products.
3. Contaminants, other than Lim, from food and other products that have been in contact with the PET packaging products.

There are guidelines [20] on how to apply an EFSA challenge test to food-grade recycling processes using compounds (i.e., marker compounds) that mimic (in polarity and MW) potential contaminants in post-consumer waste (Chapter 4). However, there are no equivalent guidelines for a quality-control test on materials and products that target contaminants that are *actually* present in post-consumer PET, or other plastics that are being recycled for food contact use (e.g., high-density PE). One of the major reasons for this scenario is that the compounds indicative of the specific plastic being recycled vary considerably, due to their chemistry and its end-uses, and so the analytical conditions to detect and quantify them vary as well.

The analytical method developed by SuperCleanQ into a CEN Technical Specification was specific to PET and cannot be employed immediately for other recycled food-grade plastics – the marker compounds and the polymer matrix are different. However, because the test is based on a relatively straightforward analytical technique, it could be used as a starting template for the development of analytical methods to determine marker compounds in other plastics. The analytical method is not intended to replace some of the other quality-control tests available for rPET materials and products, but it can be used to compliment and support these tests, and the ways in which this can be achieved are described below.

7.6.2.1.1 In support of the European food safety authority 'challenge test' for food-grade polyethylene terephthalate recycling processes

Only a small sample is required for the SuperCleanQ test, so it can be used to obtain 'benchmark' data on a PET recycling process and can be employed as and when required to determine if the level of the six compounds in the rPET has varied with time which, in itself, could indicate that the process has altered in some way. This information is valuable for industry because EU Regulation EC 282/2008 stipulates that a EFSA challenge test should be run initially to demonstrate that a recycling process is capable of decontaminating post-consumer PET to a level acceptable to the EFSA, and that it should be re-run if the company responsible for the recycling process think that anything has changed since the initial assessment. This method, therefore, provides industry with a relatively easy and cost-effective tool for providing information to determine if there is a need to re-run the more expensive and disruptive challenge test.

7.6.2.1.2 In support of the tests to assess the quality and regulatory compliance of recycled polyethylene terephthalate – in flake, granule or final product form

This method is very flexible in that it can be applied to PET samples in any physical form. It is, therefore, possible to use it on all types of PET products and their intermediates. It is also possible to use it on samples taken from any stage in the recycling process – from feedstock to the final granulate that will be provided to a converter. Because it measures the concentration of six compounds, a number of which arise due to the degradation of the PET, it is possible to use it to indicate the state of the material to establish benchmark values for 'good' or 'acceptable' samples or products. The addition of this analytical method into the suite of quality assurance and control tests available to industry (Sections 7.1 and 7.2) is, therefore, an important contribution to the rPET sector. Its use could assist in the following:
- Early warning regarding a change in the quality of the post-consumer PET feedstock for a recycling process.
- Enhanced supplier and customer confidence that food-grade PET materials and articles containing rPET are of a consistently high standard that meets regulatory requirements.

7.6.2.2 Method for detecting contaminants in recycled polyethylene processing lines by inline monitoring

In addition to the analytical method described above, one of the other objectives of the SuperCleanQ research project was to develop an inline monitoring method for the detection of contaminants during the processing of rPET. The ability to detect contaminants in processing lines using an inline monitoring system enables real-time quality control and provides an organisation with the following:
a) Faster response to deviations in process conditions.
b) Reduction in manufacturing losses due to non-compliance of products.

The development work carried out during the course of the project resulted in the development of an inline monitoring system based on an near-infrared spectroscopy

(NIR) detector capable of detecting two contaminants associated with biodegradable packaging: polylactic acid (PLA) and oxobiodegradable additives (i.e., metal catalyst-based additives). It was able to detect PLA in the PET melt at ≤0.01% and oxobiodegradable additives at 0.1%.

The potential problems that can be associated with these two contaminants in plastic recycling streams are discussed in Section 3.5. The importance of being able to detect them during processing is highlighted by the strong growth in biodegradable packaging over the last 5 years and the high predicted growth rate for the rest of this decade. As a consequence of this growth, due to the increasing amount of biodegradable packaging that will be present in the market, even if effective steps are taken to segregate it from the conventional plastic waste stream, some could still become mixed up with it and so the ability to detect these types of contaminates will continue to be of commercial importance.

As well as work to develop the inline NIR detection system, the SuperCleanQ project, also carried out work to determine the influence that PLA and oxobiodegradable additives had on the properties of rPET. Injection-moulded test pieces were prepared from PET that had been doped with these contaminants and these were tensile tested to see the effects on its mechanical properties. In the concentration range studied, there was no significant effect of either contaminant on the strength or elasticity of PET, although as little as 0.05% PLA affected the visual clarity of PET. Thermal analysis (DSC) was used to study the effect of the contaminants on crystallinity and melting behaviour. In the concentration range studied, neither contaminant had any significant effect on the melting point, but the crystallinity was increased slightly by 2% of PLA or oxobiodegradable additive, though this had little effect on the mechanical strength of the PET. The data also indicated that there was little degradative effect of either contaminant within the concentration range that the SuperCleanQ inline NIR system was capable of detecting.

References

1. *Develop a Post-Market Test for Recycled Food Contact Materials*, Food Standards Agency Project FS241007, Final Report, Food Standards Agency, London, UK, 11th March 2014.
2. CEN/TR 15353: Plastics – Recycled Plastics – Guidelines for the Development of Standards for Recycled Plastics, European Committee for Standardization, Brussels, Belgium, 2007.
3. M.J. Forrest in *Analysis of Plastics*, Rapra Review Report No.149, Smithers Rapra Technology Ltd, Shawbury, Shropshire, UK, 2002.
4. R. Brown in *Handbook of Plastic Test Methods*, John Wiley & Sons Inc., New York, NY, USA, 1989.
5. ISO 1133: Plastics – Determination of Melt Flow Rate, International Organization for Standardization, Geneva, Switzerland, 2011.
6. ISO 1628-5: Measurement of the Inherent Viscosity of PET, International Organization for Standardization, Geneva, Switzerland, 1998.
7. ISO 16014: Determination of Molecular Weight by Gel Permeation Chromatography, International Organization for Standardization, Geneva, Switzerland, 2012.

8.   ISO 180: Plastics – Determination of Izod Impact Strength, International Organization for Standardization, Geneva, Switzerland, 2000.

9.   ISO 527-1 and -2: Plastics – Determination of tensile properties, International Organization for Standardization, Geneva, Switzerland,

10. ASTM F2013-10: Determination of Residual Acetaldehyde in PET Bottle Polymers, American Society for Testing and Materials, Washington, DC, USA, 2010.

11. W. Ramao, M.F. Franco, M.I.M.S. Bueno and M.A. De Paoli, *Polymer Testing*, 2010, **29**, 7, 879.

12. W. Ramao, M.F. Franco, A.H. Iglesias, G.B. Sanvido, D.A. Maretto, F.C. Gozzo, R.J. Poppi, M.N. Eberlin and M.A. De Paoli, *Polymer Degradation and Stability*, 2010, **95**, 4, 666.

13. Anon, *SA Plastics Composites and Rubber*, 2014, **11**, 6, 78.

14. A. Ptieek Sirocic, L. Kratofil Krehula, Z. Katancic, A. Reseek, Z. Hrnjak Murgic and J. Jelencic, *Chemistry in Industry*, 2011, **60**, 7–8, 379.

15. M. Gneuss, *International Fiber Journal*, 2010, **25**, 3, 44.

16. Anon, *PETplanet Insider*, 2012, **13**, 3, 1438.

17. K.B. Prabhu, G.J. Chomal and S.M. Kulkarni in *Proceedings of the SAMPE 2013 Conference*, Long Beach, CA, USA, 6–9th May 2013, Ed., SAMPE International Business Office, Covina, CA, USA, 2013, Paper 98, p.15.

18. *Guidance and Criteria for Safe Recycling of Post Consumer Polyethylene Terephthalate into New Food Packaging Applications*, Report EUR 21155, Office for Official Publications of the European Communities, Luxemburg, 2004.

19. F. Welle, *Resources, Conservation and Recycling*, 2013, **73**, 1, 41.

20. *EFSA Journal*, 2011, **9**, 7, 2184.

21. EN 13130 (Parts 1 to 28): Materials and Articles in Contact with Foodstuffs – Plastics Substances Subject to Limitation, European Norms, Brussels, Belgium, 2004.

22. M. Whitt, K. Vorst, W. Brown, S. Baker and L. Gorman, *Journal of Plastic Film and Sheeting*, 2013, **29**, 2, 163.

23. G.G. Shimamoto, B. Kazitoris, L.F.R. De Lima, N.D. De Abreu, V.T. Izabel Maria E. Salvador, M.S. Bueno, E.V.R. De Castro, E.A.S. Romao Wanderson and E. Filho, *Quimica Nova*, 2011, **34**, 8, 1389.

24. W. Ramao, M.F. Franco, M.I.M.S. Bueno, M.N. Eberlin and M.A. De Paoli, *Journal of Applied Polymer Science*, 2010, **117**, 5, 2993.

25. R. Franz, A. Mauer and F. Welle, *Food Additives and Contaminants*, 2004, **21**, 3, 265.

26. F. Welle, *Food Additives and Contaminants*, 2008, **25**, 1, 123.

27. Y. Ishida, K. Ohsugi, K. Taniguchi and H. Ohtani, *Analytical Sciences*, 2011, **27**, 10, 1053.

28. J.D. Badia, E. Stromberg, S. Karlsson and A. Ribes-Greus, *Polymer Degradation and Stability*, 2012, **97**, 1, 98.

29. R. Franz, *Food Additives and Contaminants*, 2002, **19**, Supplement, 93.

30. C. Nerìn, J. Albiñana, M.R. Philo, L. Castle, B. Raffael and C. Simoneau, *Food Additives and Contaminants*, 2003, **20**, 7, 668.

31. F.L. Bayer, *Food Additives and Contaminants*, 2002, **19**, Supplement, 111.

32. S.D. Mancini, A.R. Nogueira, J.A.S. Schwartzman and D.A. Kagohara, *International Journal of Polymer Materials*, 2010, **59**, 6, 407.

33. F.L. Bayer, D.V. Myers and M.J. Gage in *Proceedings of the 208th American Chemical Society National Meeting*, American Chemical Society, Washington, DC, USA, 25th August, 1994, p.152.

34. *PET Recyclability 'Guidelines and Criteria for Safe Recycling of Post-Consumer PET into New Food Packaging Applications'*, EU Project FAIR-CT98-4318: Section I.

35. H. Widen, A. Leufven and T. Nielsen, *Food Additives and Contaminants*, 2004, **21**, 10, 993.

36. J. Jetten and J.C. Ravensberg in *Challenge Test of the PTP modified PET-M Recycling System*, TNO V6633 Report, 25th October 2005.

37. M. Monteiro, C. Nerin and F.G.R. Reyes, *Packaging Technology and Science*, 1999, **12**, 241.
38. K. Hinrichs and O. Piringer in *Evaluation of Migration Models*, Final Report, Thematic Network EU Contact SMT4-CT98-7513, 2001.
39. M. Mutsuga and Y. Kawamura, *Food Additives and Contaminants*, 2006, **23**, 2, 212.
40. *An Investigation of Functional Barriers Currently used by the Food Industry and an Assessment of their Efficiency*, Final Report, FSA Project A03049, Food Standards Agency, London, UK, March 2009.
41. *Large-Scale Demonstration of Viability of Recycled PET (rPET) in Retail Packaging*, Waste and Resources Action Programme (WRAP), Banbury, UK, 21st June 2006.
42. N. Dimitrov, L. Kratofil Krehula, A. Pticek Sirocic and Z. Hrnjak-Murgic, *Polymer Degradation and Stability*, 2013, **98**, 5, 972.
43. SuperCleanQ EU Funded FP7 Research Project. http://www.supercleanq.eu
44. CEN/TS 16861: Plastics. Recycled Plastics – Determination of Selected Marker Compounds in Food Grade Recycled Polyethylene Terephthalate (PET), European Committee for Standardization, Brussels, Belgium, 2015.

8 Food-grade products made from recycled polyethylene terephthalate

8.1 Introduction

This section provides an overview of the principal products that can be manufactured using the food-grade recycled polyethylene terephthalate (rPET) resulting from use of the mechanical or chemical recycling processes and routes covered in Chapter 6. As would be expected, most of the products included in this section are food contact products (e.g., bottles), but food-grade rPET can be used for packaging other products where it is desirable to have a pure rPET container that will not contaminate the packaged product (e.g., mouthwash and shampoo), and these are covered as well.

One of the most common uses for polyethylene terephthalate (PET) and, as a consequence, rPET, is in the manufacture of food packaging, and PET is the material of choice for bottles and a large proportion of non-bottle rigid packaging, such as thermoformed trays. Because this is one of the major markets for rPET, there are numerous references to food-grade rPET and its use to produce beverage bottles, food trays and other food contact products throughout this book (e.g., Chapters 2, 3, 5 and 7). The information provided in this section can, therefore, be regarded as being complemented by the references provided in these other sections and, together, they provide a fuller understanding of the food-grade rPET industry.

The fact that many recycling processes have been developed for the production of food-grade rPET, and that many applications have been made to the European Food Safety Authority (EFSA) to have these processes assessed, has been described in Chapters 4 and 6. This high level of activity also highlights the significance of the food-packaging sector to the PET recycling industry. In addition to its specific importance to PET, the packaging of food is of major importance to the plastics sector in general. This fact has been highlighted by recent data produced by Mergers Alliance in 2012 [1] showing that of the 57 million tonnes of plastic produced in Europe each year, ≈46 million tonnes are converted into products and, of this, ≈18 million tonnes is made into packaging. The market split for this ≈18 million tonnes of packaging is shown in Table 2.8.

In the specific case of rPET, the future for marketing and selling rPET for food packaging is believed to be secure [2]. Due to high demand, in recent times food-grade rPET has been almost equal in price to virgin food-grade PET but, as the number of food-grade rPET facilities increases, this will reduce cost and the attractiveness of the recycled alternative will be enhanced. To support and grow this market, there continues to be a healthy amount of research in areas associated with using recycled plastics in food contact products, including use of rPET as well as several other

https://doi.org/10.1515/9783110640304-008

plastic types. For example, a recent review of the progress of the research in identification technologies for food-related plastics products containing recycled material was published by Lu and co-workers [3]. Also, there have been UK and European Union (EU) government-funded research projects in this area. Examples include the SuperCleanQ Framework 7 EU Project [4], a UK Foods Standards Agency research project [5], and several projects funded by Waste and Resources Action Programme (WRAP) in the UK looking at aspects of recycling, such as sorting and identification [6], and the re-use of rPET in food packaging [7, 8].

An example of the work that was carried out by WRAP as part of the 'Improving food-grade rPET quality for use in UK packaging' project [8] is shown in Figure 8.1. This photograph is taken from the part of the research that looked to improve the colour and clarity of rPET and shows examples of the colours of preforms and bottles made with 50% rPET with additions of toners and optical brighteners.

Figure 8.1: Examples of the colours of preforms and bottles made with 50% rPET and the addition of toners and optical brighteners to improve colour and clarity. Reproduced with permission from the Waste and Resources Action Programme, Banbury, UK. ©WRAP.

There are many commercially available recycling systems for the production of food-grade rPET, as shown in Chapter 6. However, some of these systems have achieved much greater market penetration that others. It is claimed that the

Vacurema technology owned by Erema is the clear global market leader with a 50% share in the rPET direct food contact market [9]. For example, in the US, of the 779,000 tonnes of PET collected each year, 295,000 tonnes are processed into rPET for direct food contact, and 50% of this amount is processed using Vacurema technology.

For many years, the only way of obtaining rPET for the manufacture of food contact products was to obtain the material from a producer using an approved recycling process [e.g., approved by the EFSA or US Food and Drug Administration (FDA)] for the treatment of post-consumer waste. Then, this rPET would often be blended with virgin polyethylene terephthalate (vPET) to produce food contact articles. Within the last few years, however, another route has opened up, with PET resin suppliers producing grades that are combinations of rPET and vPET. Examples of this are the food-approved grades produced by leading PET resin producer La Seda de Barcelona, designated Artenius Unique and Artenius Elite. Artenius Unique is available in nine grades containing 10, 25 or 50% rPET, and the Artenius Elite resin has been reported to contain 50% rPET. They are manufactured using a proprietary chemical recycling process [10] that takes transparent and light-blue rPET, decontaminants it and then converts it to rPET flakes. The latter are then depolymerised using a glycolysis process to enable it to be used as feedstock, along with vPET raw materials, in the polymerisation process. Any products made from these resins can be collected in the same way as products produced from 100% vPET and recycled. Other resin suppliers that are launching food-grade rPET products include Phoenix Technologies, who in 2013 announced LNOTM w resin, a melt-extruded rPET pellet for ambient/cold/frozen food applications [11], which can be used in food-packaging products at ≤100%, or blended with vPET.

To be used for food contact packaging, rPET must meet EU food contact regulations [e.g., the Plastics Recycling Regulation EC 282/2008 and the Plastics Regulation EU 10/2011 (Chapter 4)]. However, this is a minimum requirement as far as certain sectors of the food industry is concerned, and several brand owners are much stricter in the criteria they use in their decision-making. Not all commercial PET recycling processes that meet the regulatory requirements will be acceptable to them [12]. This fact needs to be borne in mind when assessing the commercial potential for any new PET recycling process as in reality it will depend upon its acceptance by the food industry as well as its technical capabilities.

It was reported in *PETplanet Insider* [13] that APPE (the Packaging Division of La Seda de Barcelona, Spain) stated that the lack of high-quality food-grade rPET was hampering the aspirations of brand owners and product fillers to increase the amount of rPET in their packaging. This concern regarding quality, as well as limitations on the available supply of food-grade rPET, has also been raised in other sections of the trade press [14]. In this article, it was stated that concerns in these areas were tempering expectations of dramatic growth for rPET in food packaging, despite commitments by Coca-Cola and Evian Volvic to use the material.

For several years it has been the replacement of glass that has been a driver for growth in the use of PET for food packaging. However, developed countries have now largely exhausted this potential and so attention is shifting to developing countries, where PET can be used as a good barrier packaging for beer, wine, juices and dairy products [15]. Alongside this trend for the replacement of glass by PET, there are other trends that have been commented on which indicate a move in the other direction, away from PET. For example, an article in *Polymer Society* [16] reviews the use of stretch blow-moulding for the production of polypropylene (PP) bottles which, due to biaxial orientation enhances physical properties such as tensile strength, clarity and gas barriers. Such bottles can be used to replace PET, polyvinyl chloride and glass packaging in a wide range of food and non-food containers.

In addition to bottles, there are also opportunities within other types of food packaging to create products manufactured using 100% rPET. An example of this situation was revealed when Invicta Plastics claimed in 2013 that they had become the first company to create the first rigid, food-safe products (e.g., cups) from 100% rPET [17]. It was reported that this development was attracting the interest of companies such as Coca-Cola Enterprises, Greenpac and Asda. Phoenix Technologies announced in 2013 that they had launched LNO^{TM} resin, a new food-grade rPET pellet that could be used at ≤100% for the manufacture of ambient/cold/frozen food packaging, but that it was not being intended to be used for packages subjected to heat during filling or during consumer use. The material, in common with other food contact grades of rPET, also has a wider potential market because it can be used for personal care packaging applications which can benefit from food-grade rPET attributes which align with corporate philosophy and marketing objectives [11].

PET is used widely throughout the world for the production of beverage bottles, including carbonated soft drinks (CSD), mineral water, ready-to-drink tea (e.g., iced tea and green tea), juices and energy drinks. It is also used in the manufacture of bottles and jars for food, detergents, cosmetics and pharmaceuticals and in thermoforming applications to make products such as food trays. The factors that drive PET demand in each end-use sector are influenced by the alternative packaging materials and pack types that PET is competing with in each segment of the market. Information on the packaging materials (e.g., glass, metal, cartons and other plastics, such as PP) that complete with PET for food packaging is shown in Table 2.14. The fast growth in the use of PET in the areas illustrated in Table 2.14 is apparent by that fact that in 2012 PET alone accounted for 61% of the global soft drinks market, mainly at the expense of glass. Some of the advantages of PET in this market are that PET bottles do not break and can be resealed and advances in technology (e.g., barrier layers, oxygen scavengers, ultraviolet (UV) blockers) have addressed shelf-life and other issues, which has also assisted PET in its growth. The information presented in Table 2.15 confirms that the PET packaging market is very buoyant and that all sectors are predicted to grow at an increasing rate in the immediate term. The market for food-grade rPET is, therefore, expected to be strong due to the increasing interest in using PET for all packaging applications and the principal types of food packaging that are manufactured from rPET are described in Sections 8.2–8.7.

8.2 Bottles

Mention has already been made of the significant number of PET bottles collected and recycled (Chapters 2 and 3). This, together with the rapidly growing infrastructure for recycling this post-consumer product into food-grade rPET has ensured that there is sufficient material available to produce new bottles and so achieve bottle-to-bottle recycling.

In Section 8.1, the research work that has been carried out by WRAP was introduced and, in one of their projects, '*Large-scale demonstration of viability of rPET in retail packaging*' [7], they collaborated with two large retailers in the UK, Marks & Spencer, and Boots. During the work that they carried out with Marks & Spencer, square juice bottles were produced that were a blend of 30% food-grade rPET and 70% food-grade vPET. Examples of these products are shown in Figure 8.2.

Figure 8.2: Marks & Spencer square juice bottles with 30% food-grade rPET content. Reproduced with permission from the Waste and Resources Action Programme, Banbury, UK. ©WRAP.

Bottle-to-bottle recycling is well established in the US, where the FDA has described acceptable approaches for the use of rPET in bottle applications. These involve the use of a surrogate compound testing procedure (i.e., challenge testing) for the recycling process, which is closely related to the EFSA test for marker compounds used in the EU. A Smithers PIRA report [18] stated that the recycling rate for PET containers in the US increased 7 years in a row to stand at 29% in 2010 and that the amount of rPET re-used was also at its highest value (454,000 tonnes). Applications for food and drink represented 21.6% of this value. This report also stated that >50% of the PET collected in the US in this period was exported to China.

With respect to approvals for new food-grade PET recycling processes in the USA, a process for producing PET food packaging from 100% rPET bottle flakes has been given approval by the FDA. This process was developed by Gneuss using a combination of a multiple-rotation extruder, melt-filtration system, and a solid-state polymerisation (SSP) process to increase the viscosity of the final PET product to the required level [19]. The process is claimed to have several advantages over conventional recycling processes: relatively low investment, low energy costs and enhanced product quality. The FDA approval is for the pellets from the process to be used for the production of hot- and cold-filled PET bottles. In addition to this example, several other 'letters of no objection' have been issued in recent years by the FDA for recycling processes intended to produce food-grade rPET (Chapter 4).

Another aspect of the food-grade rPET market concerns the setting up of 'in-house' systems within large food manufacturing factories. These set ups, in which bottles containing rPET are manufactured on the same site where the food product is manufactured, are economically and environmentally attractive because they remove the need to transport large numbers of empty bottles to a site for filling. An example is the in-house bottle-manufacturing facility at the GSK Lucozade and Ribena factory in Coleford, Gloucestershire, in the UK. Logoplaste operate the facility that converted over 15,000 tonnes of vPET and 4,000 tonnes or rPET into bottles in 2012 [20].

Treece and Pecorini [21] have reviewed the progress made over the last three decades in extrusion blow-moulding (EBM) for the production of PET bottles. The authors state that the optimal formulation for a clear EBM material must fulfill three main requirements: process efficiently on existing equipment, produce bottles with robust drop performance, and have an acceptable recyclability story. They also state that balancing all three criteria in a single formulation is challenging because obtaining compatibility in the PET recycling stream inherently causes drawbacks to processing and bottle performance. However, the authors conclude that the significant degree of innovation that has taken place over the last 30 years has provided a comprehensive portfolio of technologies suitable for various requirements.

It was reported by Yuan and co-workers [22] that it is now common to see between 25% and 30% rPET in CSD bottles but, that as the availability of food-grade rPET increases, owners of food and beverage brands are pushing for higher rPET content in their packaging. These researchers have carried out a study to assess if higher levels of rPET reduce the physical properties of the material and so hinder its use for the manufacture of pressurised bottles. They quantified changes in the short and long-term properties that govern the ability of a bottle to retain its shape if subjected to sustained carbonation pressurisation. Commercially available bottles containing rPET at two levels (30% and 100%) were examined and the results of tensile tests showed that the 100% rPET bottle was stiffer and tougher in the axial direction (≤26%) but softer and weaker in the hoop direction (≤14%) compared with the 30% rPET bottle. The creep tests also showed that the 100% rPET sample creeped 50% faster. The workers concluded that failure to retain shape adequately would affect important properties, such as stacking stability and vending performance. Obviously there are several ways in which food-grade rPET can be produced (Chapter 6) and such work must be repeated on rPET from a number of these ways to assess how representative these findings are overall because there have been reports of 100% rPET CSD bottles being introduced into the market. For example, an article in *Plastics and Rubber Asia* announced in 2012 [23] that PepsiCo has introduced the first soft drinks bottle, called 7UP EcoGreen, made from 100% rPET, in the US. This product has also been the subject of an article in *European Plastics News* [24] where it is mentioned that this bottle was intended for diet and regular versions sold in Canada and was expected to reduce the amount of vPET used for those products by 6 million pounds per year. There are also other PET bottles in commercial use for carbonated drinks that contain >40% rPET. An example is a 1.5-l bottle for Lidl. In addition to being relatively high in rPET content, this bottle is also one of the lightest bottles in the world at 26.8 g [25].

In addition to the use of increasing amounts of rPET in PET bottles, there has also been a trend to reducing the weight of PET bottles and so reduce the amount of PET that ends up in the waste stream at the end of its life. For example, PET bottles used by the company Lidl (see above) in its own Freeway and Saskia brands are manufactured using >40% of rPET and it has been shown to be possible to reduce the weight of these bottles by ≈30% while maintaining their functionality and handling qualities [25].

The ability to recycle PET bottles more than once is an attractive proposition, but there is an obvious need to fully assess the effects of this on several important properties. Such an investigation has been carried out by a group of researchers from Cologne and Kassel Universities in Germany [26]. The team developed a rigorous process model which described a closed-loop recycling system for PET beverage bottles and targeted the dominant quality parameters such as intrinsic viscosity, acetaldehyde concentration, concentration of carboxylic end-groups and concentration of vinyl end groups. The model covered the main process steps of:
- Preform production by injection-moulding
- Drying
- SSP
- Melt filtration

The simulation that they carried out revealed that after a single recycling loop, all the relevant quality parameters achieved specification, if certain temperatures, residence times, and surface areas for degassing are provided. Another simulation was carried out in which an 'infinite' number of recycling loops were performed on PET in a closed system, and this showed the following effects on quality parameters:

- The concentration of acetaldehyde and vinyl end groups decreased with the number of recycling loops.
- The concentration of carboxylic end groups increased with each completed recycling loop, which made the PET more susceptible to hydrolysis and increased the SSP process time needed to achieve the required intrinsic viscosity for CSD bottles.

The workers suggested that to overcome the intrinsic viscosity problem, the rPET could be blended with vPET.

Petainer has developed a refillable PET beverage bottle that it regards to be its greenest ever according to an article in *British Plastics and Rubber* [27]. The new bottle contains >25% of rPET, but is claimed to have the same performance characteristics as a refillable bottle made entirely from vPET.

8.3 Large drinks containers

The company Lightweight Containers announced in 2013 [28] the launch of KeyKeg 20, a cylindrical container developed by them in response to customer demand for one-way, lightweight packaging that could be used for highly carbonated beverages. An innovative feature of the container is called 'double-wall technology', which results in a 'cylinder within a cylinder' structure imparting resistance to higher internal pressures with low deformation. The shell of the KeyKeg contains 54% rPET and the product has an ergonomically designed grip made from 100% recycled PP. The container is claimed by the company to be robust enough to withstand hot climates and long journeys, is resistant to moisture and logistically effective.

8.4 Trays and bowls

Section 8.2 highlighted the work carried out by WRAP during their '*Large-scale demonstration of viability of rPET in retail packaging*' project [7]. In addition to the work that they carried out with Marks & Spencer relating to square juice bottles (Figure 8.2), salad bowls were produced using a thermoforming process from PET containing 50% food-grade rPET. Examples of the products produced are shown in Figure 8.3.

Figure 8.3: Producing Marks & Spencer salad bowls with 50% food-grade rPET content. Reproduced with permission from the Waste and Resources Action Programme, Banbury, UK. ©WRAP.

Improvements in the aesthetic appearance and performance of packaging trays has been shown to be beneficial in increasing sales according to an article published in *Food Packer and Processor International* [29]. The article explains how collaboration between Linpac Packaging Mondini, and the film manufacturer Bemis has resulted in the creation of a lightweight, shallow, rigid tray for vacuum skin packing that can be produced from rPET, PP, or expanded polystyrene. Use of this tray, which gives a better quality look and feel to the packs and extends the shelf-life of meat products, is claimed to have boosted the meat product sales of Booths, a family-owned supermarket chain, by 80%.

8.5 Film and sheet

A group in the US [30] have characterised 60 extruded sheets of PET that contained between 0% and 100% rPET using spectroscopic techniques [inductively coupled plasma-atomic emission spectroscopy (ICP-AES) and UV–visible], thermal analysis by differential scanning calorimetry, and mechanical tests. Their results showed that absorption at 350 nm, percentage crystallinity and the crystallisation peak offset were good indicators of the level of rPET. The results of the mechanical tests showed that incorporating rPET into virgin resin significantly altered the properties of the resulting products. For example, at 40% rPET, in the machine direction,

Figure 8.4: Boots 'ingredients' products in bottles with 30% rPET content. Reproduced with permission from the Waste and Resources Action Programme, Banbury, UK. ©WRAP.

there was a 2-MPa increase in stress at the proportional limit, a 3-MPa increase in the stress at yield, and a 30-MPa increase in Young's modulus compared with 100% vPET. The ICP-AES results showed that the rPET/PET sheets could be used safely for food packaging according to the California Health and Safety Code.

8.6 Use of non-food-grade recycled polyethylene terephthalate in food contact products

Manufacturers often wish to use rPET in their food contact products. However, either due to cost or availability, they choose not to use food-grade rPET, but instead non-food-grade rPET. Also, they use it in the product for surfaces that do not directly contact the food, although care must be taken in using this option because the EU regulations also cover plastics that 'may come into contact with food in typical use' (Chapter 4). Another possibility is to use the non-food-grade rPET in a 'sandwich' construction between layers of food-grade vPET. The sandwich construction tends to be used the most, and an example of food packaging produced from a vPET/ rPET/ vPET sandwich is the Rfresh line from Linpac Packaging. This packaging can be used to store food products, such as fresh meat, fish and poultry, for ≤10 days in standard chiller distribution systems. Linpac have processing systems that are capable of using post-consumer waste and its own industrial waste [31].

Packaging design and technology continues to advance and produce highly technical products that can meet the technical demands of manufacturers and their environmental goals. An example is a plastic beer bottle produced by Owens-Illinois that comprises five layers. The bottle is composed of vPET inner and outer layers, a centre layer of rPET, and layers of proprietary SurShield Nylon-based barrier material sandwiched between both of the vPET layers and the rPET layer [32].

8.7 Cosmetics and healthcare products

Cosmetics and healthcare products include some readily identifiable products, such as shower gel, mouthwash, shampoo, skin creams and cleansers, and bath lotions. Many of these products are packaged in containers (e.g., bottles) for which PET is an excellent material. Providing the properties, particularly mechanical properties, are sufficiently good there is no reason why rPET cannot be used in their manufacture. It could be argued that this section should have been incorporated into Chapter 9 of this book which deals with 'other products manufactured from rPET'. However, because the products that the PET packaging contains often come into intimate contact with human skin (e.g., mouthwash) the PET that is used for their manufacture (vPET or rPET) is usually food-grade.

The use of rPET for the manufacture of these types of products has been encouraged in the UK by research projects funded by the government recycling agency WRAP. For example, as part of one of WRAP's projects, Boots incorporated 30% rPET into the packaging it was using for its Ingredients range of haircare products. It was also reported that, as part of the same WRAP research project, Coca-Cola tested different types of rPET, such as pellet and flake forms, in its packaging at an inclusion rate of 25% [33].

References

1. Mergers Alliance, Plastics Europe Market Research Group (MRG), Rexam, 2012. http://www.mergers-alliance.com
2. Anon, *Bioplastics World*, 2013, **1**, 10, 4.
3. S. Lu, Z. Xiongyan, S. Zhanying, Z. Shibiao, X. Jijun, M. Jinsong, W. Xin, W. Lei, L. Hongbin, Q. Wanbao, H. Guang and D. Guofa, *Plastics Science and Technology*, 2013, **41**, 12, 91.
4. SuperCleanQ EU Funded FP7 Research Project. http://www.supercleanq.eu.
5. *Develop a Post-Market Test for Recycled Food Contact Materials*, Project FS241007, Final Report, Food Standards Agency, London, UK, 11th March 2014.
6. *Development of NIR Detectable Black Plastic Packaging*, Final Report, Waste and Resources Action Programme (WRAP), Banbury, UK, September 2011.
7. *Large-Scale Demonstration of Viability of rPET in Retail Packaging*, Final Report, Waste and Resources Action Programme (WRAP), Banbury, UK, June 2006.

8.   *Improving Food Grade rPET Quality for use in UK Packaging*, Final Report, Waste and Resources Action Programme (WRAP), Banbury, UK, July 2013.

9.   Anon, *Popular Plastics and Packaging*, 2014, **59**, 3, 53.

10. Anon, *European Plastics News*, 2012, **39**, 9, 42.

11. Anon, *PETplanet Insider*, 2013, **14**, 6, 24.

12. Anon, *PETplanet Insider*, 2011, **12**, 3, 20.

13. Anon, *PETplanet Insider*, 2012, **13**, 3, 10.

14. Anon, *Advanced Packaging Technology World*, 2013, **1**, 6, 2.

15. F. Welle, *Kunststoffe International*, 2013, **103**, 10, 64.

16. L.R. Subbaraman, *Polymer Society*, 2011, **4**, 2, 41.

17. Anon, *British Plastics and Rubber*, 2013, March, 9.

18. D. Platt in *Future of Global PET Packaging to 2017*, Smithers PIRA, Leatherhead, UK, 2012.

19. Anon, *High Performance Plastics*, 2010, April, 11.

20. Anon, *Plastics and Rubber Weekly*, 2013, 4th October, 23.

21. M.A. Treece and T.J. Pecorini in *Proceedings of the 70th SPE Annual Technical Conference*, Orlando, FL, USA, 2–4th April, Ed., Society of Plastics Engineers, Brookfield, CT, USA, 2012, Paper No.7.

22. J.Z. Yuan, C.A. Haynes and P.A. Harrell in *Proceedings of the 71st SPE Annual Technical Conference*, Cincinnati, OH, USA, 22–24th April 2013, Ed., Society of Plastics Engineers, Brookfield, CT, USA, 2013, Paper 1584107.

23. Anon, *Plastics and Rubber Asia*, 2012, **27**, 187, 18.

24. Anon, *European Plastics News*, 2011, **38**, 9, 34.

25. Anon, *PETplanet Insider*, 2012, **13**, 1–2. 26.

26. T. Rieckmann, F. Frei and S. Volker, *Macromolecular Symposia*, 2011, **302**, 34.

27. Anon, *British Plastics and Rubber*, 2012, February, 20.

28. Anon, *PETplanet Insider*, 2013, **14**, 4, 32.

29. Anon, *Food Packer and Processor International*, 2013, **28**, 5, 44.

30. G. Curtzwiler, K. Vorst, J.E. Danes, R. Auras and J. Singh, *Journal of Plastic Film and Sheeting*, 2011, **27**, 1–2, 65.

31. K. Coyne, *Plastics and Rubber Weekly*, 2008, 25th July, 1.

32. Anon, *Packaging Digest*, 2002, **39**, 12, 4.

33. D. Eldridge, *Plastics and Rubber Weekly*, 2005, 11th November, 3.

9 Other products manufactured from recycled polyethylene terephthalate

9.1 Introduction

Chapter 8 provided an overview of the types of products, principally food contact materials and articles, which can be made from food-grade rPET. This section carries out a similar function for the wide range of products that can be made from non-food-grade recycled polyethylene terephthalate (rPET) such as fibres and strapping. It also introduces some of the completely new products (e.g., polyurethane (PU) coatings and unsaturated thermosetting polyester resins) that can be produced *via* the depolymerisation route, which is a popular route for polyethylene terephthalate (PET) due to the relative ease in which the process can be carried out (Section 6.3), and the recovery of energy *via* generation of fuel products. In this way, Chapters 8 and 9, together with the other sections in the book that mention rPET products (e.g., Chapters 2 and 3), cover the range of what can result from the use of the four classes of recycling processes (i.e., primary, secondary, tertiary and quaternary) introduced in Chapter 1. Chapter 3 provides some European and US end-market data for food and non-food rPET products in Tables 3.1 and 3.3, respectively.

9.2 Use in construction products

A team of workers from two Universities in Iran [1] investigated the effect of incorporating different levels of PET into concrete. The group substituted the sand in the concrete at three levels, 5, 10 and 15%, with processed particles of waste PET and evaluated the resulting products. In one set of studies, cubic and cylindrical specimens with different water-to-cement ratios were produced and the physical properties of the fresh concrete assessed. Work was also carried out using specimens that had been cured under standard conditions to determine their mechanical properties. Results indicated that fresh concrete containing PET particles demonstrated a lower modulus of elasticity and splitting tensile strength compared with conventional concrete. The data also showed that although the compressive and flexural strength showed an ascending trend at the initial stages, they tended to decrease after a period of time. Ultrasonic pulse tests were also conducted and revealed that the concrete samples containing PET had a porous structure.

A company called Eco-Tec Environmental Solutions [2] has developed and introduced new house-building construction techniques using standard post-consumer PET bottles instead of conventional bricks. The building system uses standard PET bottles filled with rubble, soil, plastic debris or other materials

https://doi.org/10.1515/9783110640304-009

present on the construction site. Once filled, the PET bottle 'bricks' are linked together to form a cohesive structure using a synthetic wire mesh of material found near the site. The whole construction is then completed by applying a plaster finish to the interior and exterior surfaces of the PET 'brick' walls. By the later half of 2010, Eco-Tec Environmental Solutions had built >50 houses, water tanks and other structures using this system in Honduras, Bolivia, Columbia and India.

PETplanet Insider reported [3] that rPET from waste bottles has been used by Miniwiz to manufacture a building material called Polli-Brick. The product comprises 100% rPET and is claimed to be translucent, naturally insulated and durable. It can be assembled into rectangular panels or any customised shape and can be tailored to a simple modular, easy-to-install affordable cladding system or an interactive colour LED integrated animated building skin system. A total of 480,000 interlocking bricks were used for the façade of the EcoArk Pavilion (which formed the centrepiece of the 2010 Taipei International Flora Expo).

The importance of aromatic polyester polyols for the production of rigid polyurethane foams (PUF) is growing exponentially due to the increasing need for fire and smoke resistant materials for the construction industry according to Stoilkov and Knief of H&S Anlagentechnik [4]. They described how compact reactors in modular-formed equipment can produce in-house polyester polyols with high aromatic content from phthalic anhydride, terephthalic acid (TPA) or rPET. The process is considered to be safe by the authors compared with the production of polyether polyols due to the non-hazardous raw materials and ordinary pressures. In addition to improved fire resistance, the authors also claim that the use of these types of polyester polyols for the production of rigid PUF has increasing economic benefits to end-users.

9.3 Wind turbines

Malnati reported [5] that a significant amount of effort was being expended to develop smaller wind turbines (e.g., individual output, <1 MW) which, if used in sufficient numbers, can be just as effective as larger versions in harvesting energy from wind. A smaller size is stated as being advantageous in several areas, such as the granting of installation permits, to improvement of take-up rates due to reduced cost. Enviro-Energies Holdings has commercialised a product called a multi-vaned, magnetic levitation-based vertical-axis wind turbine that is certified for use in winds ≤120 mph. The vanes of this product are moulded thermoplastic sandwich composites formed in a low-pressure compression-moulding process. The skins of the vanes are manufactured from rPET that originates from post-consumer bottles and the resin in the vane core is post-industrial scrap.

Stewart [6] described how new and innovative core materials are adding design freedom, cost-efficiency and other desirable qualities, such as high strength stiffness and optimised weight, to composite sandwich structures. An example of such a new product is G-PET, a recyclable PET structural foam core material developed by Gurit produced from a mixture of rPET drinks bottles and textile fibres.

The G-PET structural foam is targeted at the wind-energy market. Also mentioned in the article are other core foams for sandwiches, such as a conformable fibre-reinforced foam designed for closed-moulding (TYCORE W from Webcore Technologies) and a new 100% bio-based hexagonal honeycomb core developed by EconCore using the patented ThermHex technology in an automated continuous production of honeycomb from polylactic acid (PLA). To complete the construction of the products, the skins for the cores can be unfilled PLA or PLA-reinforced with consolidated flax.

9.4 Automotive products

An article in the journal *International Fiber Journal* [7] discussed recent changes in the automotive sector with respect to the use of recyclable and recycled non-woven fabrics for vehicle interiors. Articles such as this place the recycling of PET into the general-sustainability context for a particular industrial sector. Examples of the developments, both PET- and non-PET related, covered in the article include:
- Vehicle interiors manufactured using Unifi's Repreve fibre made from a hybrid blend of recycled materials, such as post-consumer PET bottles and post-industrial waste for the Ford Focus electric car.
- Recycled polyester automotive carpets.
- Plant-based bio-PET for various interior applications.
- Plant-based bio-polyethylene (PE) for floor mats.
- Ecobond glass-free headliner.
- Fasertec coconut fibre non-wovens for use in natural rubber (NR) latex-bonded composites for seat backs and head rests.

9.5 Clothing products, fabrics and fibres

The recycling of post-consumer PET products (e.g., bottles and trays) into textile fibres has become environmentally and commercially attractive. However, rPET yarns can have inferior mechanical properties and long-term degradation resistance due to contamination within them that originates from the non-PET fraction left in the recycling stream (e.g., other polymers, and label adhesive, ink and coating residues). A group of researchers drawn from various institutions in Korea [8] carried out an in-depth study in this area which investigated the influence that different recycling processes (mechanical and chemical) have on the physical properties, thermal characteristics, resistance to hydrolysis and photo-degradation of rPET fibres. Results showed that high-purity yarns produced using chemical recycling had similar processability, physical, mechanical and ageing properties, to virgin polyethylene terephthalate (vPET) yarns.

Research work has also been carried out on the use of blends of vPET and rPET for the manufacture of textile fibres [9]. Blends containing 5–20% of rPET were prepared and their rheological characteristics studied at three temperatures (270, 275 and 280 °C) and shear stresses at 6.2–14 × 103 Pa. When the data obtained was used to draw flow curves for the melts, the researchers found that it obeyed the power law for each of the three temperatures. The group also investigated the effects of incorporating two stabilisers (Irganox 1076 and butylated hydroxy toluene), each at 0.5%, on the thermooxidative degradation of the materials. Irganox 1076 was found to be the most effective, with an increase in viscosity from 50 to 100 Pa.s being achieved. Other work carried out in the same programme included studies on mixtures at different ratios of the initial blends. Work has also been carried out by Waste and Resources Action Programme (WRAP) to produce spinning fibres from blends of rPET and vPET. These products were produced in trials during the course of the work that they sponsored to investigate the end markets for recycled detectable black PET plastics [10]. Figure 9.1 shows spinning bobbins that contain fibres produced from blends of black, detectable crystalline rPET and vPET. The fibres had levels of rPET of 0–20%.

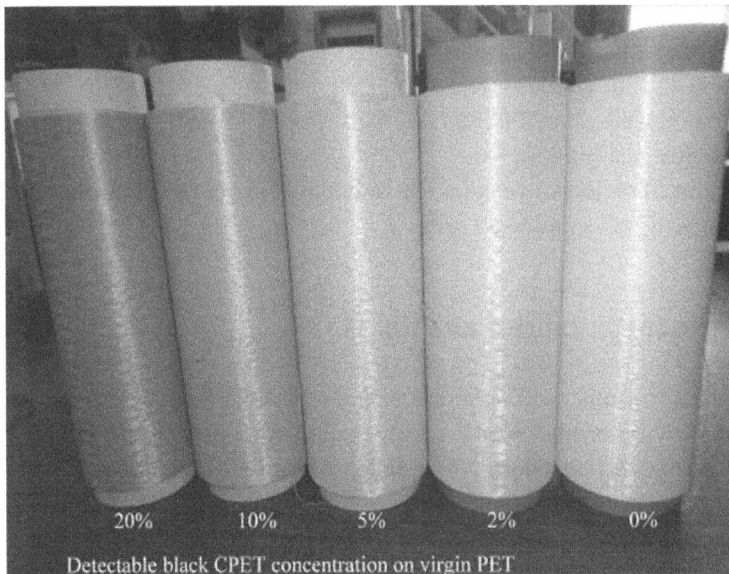

Figure 9.1: Fibres from blends of black detectable rPET and vPET (rPET content, 0–20%) on spinning bobbins produced during the WRAP trial. Reproduced with permission from the Waste and Resources Action Programme, Banbury, UK. ©WRAP.

Several clothing manufacturers are committed to using recycled plastics and the principle of sustainable design in the development of their products. One of these manufacturers is Levi and they have a line of products called 'Waste Less' in which the denim that the jeans are manufactured from contains more than just cotton. According to an article in *PETplanet Insider* [11] each pair of jeans contains ≥20% of post-consumer PET bottles and food trays, which corresponds to 12–20 fl oz bottles per pair. To enable this to take place, the bottles and trays were first sorted by colour before being crushed into flakes and converted into polyester fibre. The fibre was then blended with cotton fibre and woven with traditional cotton yarn by Cone Denim to create the fabric used in Waste Less brand clothing.

Li Di of the Tonghui Chemical Technology and Engineering Company described a recycling process, developed by this organisation, which allows rPET bottle flakes to be spun into an acceptable partially orientated yarn or non-woven material [12]. The process involves depolymerisation of the rPET and then repolymerisation of the low-molecular weight (MW) products that result to produce the final polymeric product. This combination enables complete removal of contamination present in the rPET starting material, and so enables flakes of differing quality to be used. Another advantage of the process includes good control of the intrinsic viscosity of the product.

Upasani and co-workers [13] studied the recycling of waste polyester products by partial depolymerisation to produce fibres. In their work they concentrated specifically on post-consumer PET bottles and assessed the influence that the configuration of the agitator component within the depolymerisation equipment had on the dissolution and reaction of the molten oligomeric material. They also examined the melt spinning of the polyester product formed at the end of the re-polymerisation process. Results showed that the agitator position, as well as its speed, affected depolymerisation of the rPET and that melt filtration was needed before the melt-spinning operation if a good-quality product was to be produced.

Hayes of Drexel University in the US reviewed the use of fibre-to-fibre recycling processes for recycling polyester fibres from fashion clothing [14]. One of the processes that received attention in the review was the EcoCircle fibre-to-fibre polyester recycling system. Also included were examples of commercial companies using fibre-to-fibre recycling systems to decrease the negative impact that the production of textiles and garments was having on the environment. Areas that were also covered included developments that aimed to increase the level at which clothing recycling occurred and some of the benefits and limitations for recycling polyester and production of new raw material.

Recycled plastics are being used in the manufacture of environmentally friendly non-woven fabrics for a range of applications, such as carpet backings. Shaw Industries and DAK Americas have created a joint-venture company, Clear Path Recycling, to produce rPET from post-consumer bottles. Both companies intend to use this material as a feedstock to enhance the value and sustainability of their individual product ranges [15].

9.6 Thermoset and thermoplastic polyester resins

A group at King Abdulaziz University in Saudia Arabia [16] looked to increase the number of options available for less purified grades of rPET by investigating its transformation into chemical building blocks by glycolysis to produce unsaturated polyester resins. In this work, waste PET from beverage and water bottles was glycolysed using different glycols [propylene glycol (PG), diethylene glycol (DEG), triethylene glycol (TEG)] and glycol mixtures (DEG with PG or TEG in equal amounts) in a series of experiments. The glycolysed products that resulted from these experiments were converted into unsaturated polyester resins by reacting them with maleic anhydride (MA). To create a series of final, crosslinked polyester resins, these unsaturated resins were reacted with styrene and the factors affecting the curing process investigated.

Scientists at the Iranian University of Technology in Amirkabir [17] produced rPET by glycolysis using ethylene glycol and determined the optimum glycol volume and catalyst concentration. The reaction products of the glycolysis work were then reacted with MA *via* polyesterification reactions to prepare unsaturated polyester resins. The resulting products were characterised using intrinsic viscosity, determination of hydroxyl value, nuclear magnetic resonance spectroscopy (NMR), differential scanning calorimetry (DSC) and Fourier-Transform infrared spectroscopy (FTIR).

Juraev and co-workers described how unsaturated polyethers were synthesised from alcoholysis products obtained from PET in household waste [18]. This group investigated the use of different monomers as active solvents for the unsaturated polyethers to achieve condensation reactions. They found that styrene solutions of the unsaturated polyethers were necessary for processing them into products at room temperature and acrylonitrile solutions at high temperatures were required. The effects of the concentration of initiator on the physico-mechanical properties of the crosslinked resins, as well as the time of crosslinking, were studied. It was found that increasing the initiator concentration and the crosslinking time increased the value of a number of properties, including density, impact strength, Brinell hardness, heat resistance and durability in bending.

A group of Indian researchers [19] investigated the sorption and diffusion of water in novel saturated polyester nanocomposites synthesised from glycolysed post-consumer PET products. The kinetics of sorption were studied using the equation of transport phenomena, with values of 'n' from the transport equation showing non-Fickian or pseudo-Fickian transport within the polymer matrix. The dependence of the diffusion coefficient on composition and temperature for all of the samples was also studied and, in the case of the saturated polyester samples, it was found to decrease with an increase in glycolysed PET content. The nanocomposite samples showed a smaller diffusion coefficient than the unfilled samples, which decreased with an increase in nanofiller at ≤4% wt/wt. The diffusion coefficient increased with an increase in temperature for all samples. The sorption coefficient showed a small change as the composition of the samples and the test temperature were varied, and the activation energy for diffusion and permeation was positive for all samples. The heat of sorption was also positive for all the samples, which indicated that a Henry-type mode of sorption was taking place.

Sabic have launched a new portfolio of environmentally friendly products that are moulding compounds based on polybutylene terephthalate (PBT) that has been made by the chemical regeneration of post-consumer PET. The novel products can be used in various automotive and electrical applications. Sabic claim that the manufacturing processes for this new range of PBT products require less energy and non-renewable fossil fuels compared with the manufacturing processes for conventional, fossil fuel-based materials. In a presentation to the ANTEC 2011 Conference by two Sabic employees, Rama Konduri and Fonseca, [20], a comparison of the properties of the new, environmentally friendly PBT and the conventional PBT materials was delivered.

A new unsaturated polyester urethane has been produced by Issam and co-workers [21] by reacting methylene diisocyanate with di(2,3-butenhydroxyl)terephthalate in the ratio of 1:1. The di(2,3-butenhdroxyl)terephthalate had been prepared by reacting 2 mols of cis-2-butene with 1 mol of TPA derived from the saponification of post-consumer PET bottles. The formation of the unsaturated polyester urethane as the reaction product was confirmed by the use of a CHN analyser, FTIR, 1H-NMR and ultraviolet–visible spectroscopy. The thermal properties of this new polymer were characterised by DSC and thermogravimetric analysis (TGA), and the mechanical properties were determined by tensile, elongation, hardness, adhesion and impact testing. Electrical testing to assess conductivity and electrical resistance was also carried out.

9.7 Coating products

Dias and co-workers [22] evaluated the physico-chemical and thermal properties of alkyd resin varnish, rPET from post-consumer drinks bottles, and different mixtures of these two materials. The analytical technique that the workers used to carry out this work was DSC. The authors reported that films produced from the pure varnish and the mixtures of varnish and rPET had a similar visual appearance, but that the maximum amount of rPET that could be added to the varnish without significantly altering its film properties was 2%. To calculate the kinetic parameters, activation energy and Arrhenius parameter for film samples that contained rPET at 0.5–2%, the Flynn–Wall–Ozawa (FWO) isoconversional method was used. This work showed that the presence of the larger amounts of rPET in the film samples resulted in a small change in the activation energy of the curing process. The group found that the most suitable kinetic model for describing the curing process was the autocatalytic Sestak–Berggren model.

Kathalewar and co-workers [23] employed neopentyl glycol (NPG) to depolymerise waste PET *via* a glycolysis route. The reaction was carried out using a molar ratio of 1:6 (PET:NPG) in the presence of 0.5% a zinc acetate catalyst at 200–220 °C . The progress of the glycolysis reaction was studied by determining the MW of the residual PET and by assessing the hydroxyl number of glycolysed oligomers. Once the glycolysis reaction had progressed to completion, the monomeric product was characterised using a full range of analytical techniques and then reacted with adipic acid, isophthalic acid and trimethyl propane to produce a polyester glycol. The latter was then used as a starting material, along with different commercial polyisocyanates, to manufacture a range of PU coatings, which were applied to mild steel panels. The films were evaluated for their optical, mechanical and chemical properties.

9.8 Printing paper

Teijin developed a water resistant, high wet-strength printing paper made entirely from its Ecopet recycled polyester fibre, which is derived from post-consumer PET bottles [24]. The company claim that the printing paper is highly water resistant compared with conventional wood pulp-derived paper and is not easily torn when wet, making it ideal for use in outdoor locations or wet locations. Potential applications for the paper include hazard maps, triage tags, outdoor posters, recording papers, and labels and price tags for fresh or frozen foods.

9.9 Use in adhesives

Thermosetting polymer resins tend to be used for bonding wood in the production of plywood composites. Because these types of adhesive can be regarded as an environmental problem during use due to the liberation of formaldehyde gas, investigations are underway to look for replacements and, such a programme of work, has been undertaken by Kajaks and co-workers at Riga Technical University in Latvia [25]. The aim of their work was to investigate the use of cheaper waste products, such as recycled thermoplastic polymers [e.g., polyolefins, polyamides (PA), PET] as hot melt glues for wood veneer bonding instead of these traditional thermosetting resins. To achieve this goal, recycled thermoplastic polymers from sources such as tetra-packaged waste and domestic film waste were obtained and characterised by techniques such as DSC. Bonding experiments were then done to determine how the ratio of the different polymers influenced properties such as shear strength. The data obtained showed that to reach high adhesive strength it was necessary to improve the cohesive strength of the adhesive layer. Waste PET fabric was used as reinforcement for the adhesive layer and it was found to improve the bending strength and modulus of a two-layer laminate bonded with a recycled polypropyl-ene (rPP)-based adhesive. Although exposing the adhesive bonds to water at 23 °C for 72 h was found to reduce the adhesive strength in all cases, the workers re-garded the results as demonstrating the excellent potential for using recycled ther-moplastic hot-melt adhesives for the bonding of wood.

9.10 Medical and pharmaceutical products

It was reported in *High Performance Plastics* in January 2014 [26] that researchers had transformed PET from recycled bottles into novel, small-molecule compounds that self-assemble in water into nanofibres. These nanofibres were found to be par-ticularly effective at destroying drug resistant fungi and fungal biofilms. The mech-anism involved was believed to be electrostatic interaction that enabled the nanofibres to selectively target fungal cells, penetrate their membranes, and kill them. *In vitro* studies demonstrated that the nanofibres eradicated >99.9% of *Can-dida albicans* after 1 h of incubation and no resistance to the nanofibres was found even after 11 treatments.

9.11 Use in plastic blends and composites

At a presentation to the 2012 ANTEC conference in Mumbai, India, Yatish and co-workers [27] described how it is possible to create blends using rPET with other recycled plastics such as PE, although there is a need to use compatibilisers to over-come the inherent incompatibility of the two polymers (one non-polar and the other polar) to create good-quality products.

A research group in Malaysia [28] took rPET from beverage bottles and com-pounded it with alkali-treated kenaf fibre. Acrylonitrile–butadiene–styrene (ABS) was added to the material to act as an impact modifier and to stabilise it during processing. The mechanical, thermal and morphological properties of the compo-sites were evaluated by tensile, flexural and impact testing, DSC and scanning elec-tron microscopy (SEM). The melt flow rate of the composites was also determined. Results revealed an increase in brittleness of the composites if higher percentages of kenaf fibre were incorporated into them, and also if the kenaf fibre was inter-penetrated fully into the matrix.

Workers from two universities in Thailand [29] looked into the potential of using rPET as an alternative reinforcing material to liquid crystal polymer (LCP) for *in situ* microfibrillar-reinforced composites based on high-density PE. To achieve this goal, the PE-rPET and PE-LCP composites were prepared as fibres using hot drawing pro-cesses. The group then studied the effects of draw ratios and compatibiliser [styrene–ethylene–butylene–styrene (SEBS)-*g*-MA] loading on the morphology, tensile properties, thermal stability and dynamic mechanical characteristics of the two composite systems. In as-spun samples containing compatibiliser, fibrillation of LCP domains was observed whereas the rPET domains appeared as droplets. Mechanical properties were enhanced more significantly by the compatibiliser in the case of PE-rPET samples than with PE-LCP samples. The presence of rPET in the fibres resulted in improvement in thermal stability, whereas no significant change was brought about by the use of the LCP. The researchers concluded that the results suggested that r-PET was a more effective minor-blend component than the more expensive LCP for these PE composite fibre products if good reinforcement and thermal resistance were required. Another group of Thai workers [30] were involved in a related study involving fabrication of *in situ* microfibrillar-reinforced composites of polypropylene (PP) with various proportions of rPET, and with liquid crystal copolyester and poly(*p*-hydroxybenzoic acid *co*-ethylene terephthalate), as dispersed phases. Once the samples were prepared and their rheological properties assessed, the group characterised their physico-chemical properties using a range of techniques. For example, the morphology of their fracture surfaces after testing was examined by SEM and their thermal properties, particularly the thermal decomposition temperature, determined by TGA and DSC. Results were evaluated with respect to the potential for the use of rPET as a processing aid and as a thermally stable-reinforcing agent similar to LCP.

Mondadori and co-workers [31] used melt processing to produce composites of rPET and short glass fibres with an optimised microstructure and high mechanical performance. The level of the glass fibres was varied from 0, 20, 30 and 40% *w/w* and treated with aminosilane- or epoxysilane-coupling agents. They were found to be well dispersed and bonded to the rPET matrix irrespective of what coupling agent was used and whether the rPET originated from unmodified bottle waste or from waste of this type that had been through a solid-state polymerisation (SSP) process. A high level of reinforcement was evident in the materials and the group pointed out that although the literature states that a twin-screw extruder is required to achieve this goal, they had done so using a single-screw extruder with a double-flight barrier screw. The authors reported that, in general, slightly better mechanical strengths were obtained with the SSP rPET, which could be due to its highly entangled amorphous phase arising from its higher MW.

It is important to optimise the design of plastic pellets to ensure that processibility is maximised, particularly where a blend of recycled polymers is concerned. A group of workers from the Institute of Technology in Kyoto, Japan, blended rPET with rPP and carried out a study to consider the effects of pellet geometry on the moisture absorption and thermal decomposition kinetics of the blend when it is being processed [32]. To carry out the kinetic study the group used the FWO isoconversional method. Results showed that the FWO isoconversional method was suitable for studying the thermal degradation of the blend in nitrogen, whereas the second-order polynomial function was found to fit for thermal oxidative degradation in air. Larger pellets were found to exhibit higher degradation activation energies, which suggested that they were more resistant to thermal degradation than smaller pellets. Finer pellet geometries, such as powders, were found to have higher moisture absorption due to their large surface area, although they could be dried more readily.

When it is recycled, PET can be susceptible to thermal and hydrolytic degradation that can result in poor mechanical properties and difficulties in moulding. Chinese and American scientists [33] have added several substances to rPET and determined their effect on its mouldability and mechanical properties. To enable the study to take place, chain extenders (CE), thermoplastic elastomer (TPE) and/or poly(butylene adipate-*co*-terephthalate) (PBAT) were melt-blended with rPET in a thermokinetic 'K-mixer' and then tensile test pieces produced from the blended materials by injection moulding. Several physico-chemical techniques were used to characterise the mechanical properties, rheological properties, crystallinity, and blend-phase compatibility of samples. The results enabled the authors to conclude that the use of these types of additives enabled rPET to regain its mouldability and that addition of the CE greatly enhanced its mechanical properties. It was also apparent that the rPET/TPE blends showed some improvement in mechanical properties, but that these improvements were less significant than blends with the PBAT due to the reduced level of compatibility resulting from the larger polarity differences. With the rPET blends, heat history was also found to play a part because additional annealing to increase the levels of crystallinity in the samples was found to improve their mechanical properties.

Van der Meer and Frenz of BASF [34] described how their low-MW styrene-acrylate copolymer Joncryl resins can be used as additives for thermoplastics and, by their addition to these materials, increase or decrease their viscosity. Also described is how the use of the Joncryl CE, due to their multifunctional epoxide moieties, can be used to increase the MW and melt strength of thermoplastics such as PET and rPET, PBT, PA, polycarbonates and bio-polyesters as well as their blends, during processing. The authors also described how these CE could act as reactive compatibilisers in blends of different polycondensation polymers and how the epoxide-functionalised Joncryl resins could increase the hydrolytic stability of polymers.

Workers in Japan [35] carried out an investigation to produce recycled materials with high static and impact strengths by blending rPET with ethylene-glycidyl methacylate copolymer (E-GMA) in an extruder. As the amount of E-GMA in the blend was increased initially, the team found that a gradual increase in the Izod impact strength occurred. However, if the level of E-GMA was increased to >13.5% *w/w*, a drastic increase in toughness was observed, with an attendant decrease in density. Other findings of the study were that the toughness and density of the blends was also dependent on the speed that the extruder screw rotated during the compounding operation and that, upon examination of the fracture surfaces of Izod test pieces, ductile and microporous structures were observed in the materials.

A group from the Academy of Sciences in the Czech Republic described how rPET can be toughened with a clay-compatibilised rubber-phase [36]. Their work looked into the influence of the clay content on the strength, toughness and modulus of the materials using X-ray diffraction (XRD), tensile testing, SEM and transmission electron microscopy. The best results they obtained were for a rPET material that contained a low amount of ethylene-propylene rubber that was pre-blended with the clay.

Baxi and co-workers at the National Institute of Technology in India [37] examined addition of rPET to blends comprising vPET and PBT. The modified blends were analysed by FTIR spectroscopy, SEM, XRD, and tensile and impact tests undertaken at different addition levels of the rPET. The group observed that, at the higher levels of rPET, there was a slight deterioration in tensile properties, but impact strength was increased. They found that moderate quantities of rPET gave the best results.

The presence of glass fibre-reinforcement in PET composites can present problems in recycling the product due to a potential reduction in the length of the glass fibres, with an attendant loss of physical properties. A group of researchers carried out a study to characterise a 15% *wt/wt* glass fibre-filled PET while it was being subjected to six successive recycling cycles [38]. The group tested the mechanical properties (e.g., tensile strength, elongation at break and elastic modulus) of the recycled composites. Results showed a slight decrease after each recycling cycle, which they postulated could be due to a decrease in the effective length of the glass fibre-reinforcement. By contrast, the thermal properties of the samples were not significantly affected and, overall, the authors thought that the data demonstrated that recycled glass-filled PET could be used effectively to fabricate products without affecting their mechanical performance significantly.

Ferreira and co-workers [39] carried out a study on the recycling of PET wastes from the production of non-woven fabrics, and PA wastes from old tyres. The group produced polymeric blends from these two waste streams by subjecting them to reactive extrusion in the presence of *trans*-reaction catalysts. The blends produced in this way were characterised by chemical and thermal analysis. Results indicated that *trans*-reactions had occurred between segments of polymeric chains in the two polymers, which promoted stabilisation of the products. The group concluded that the production of PET/PA blends presented an interesting alternative for the recycling of PET and PA wastes.

9.12 Use in rubber products

Powdered rPET can be employed as filler in rubber compounds. Nabil and Ismail [40] prepared NR compounds that contain powdered rPET and NR compounds where some of the rPET powder had been replaced with commercial fillers such as N550 carbon black, halloysite nanotubes (HNT) and silica. To enable them to study the effect of this partial replacement of the rPET, the two workers investigated the fatigue life, the thermal stability and the morphology of the compounds. Results showed that the partial replacement of rPET by the three commercial fillers increased the fatigue life of the NR/rPET/N550 material to a greater extent than it did with the NR/rPET/HNT or the NR/rPET/silica material. The NR/rPET/HNT compound exhibited a higher decomposition temperature than the NR/rPET/N550 and NR/rPET/silica compounds. SEM micrographs of the compounds were in good agreement with the fatigue-life results because they revealed features in the NR/rPET/N550 compounds that suggested that higher energy would be required to cause failure.

9.13 Low-molecular weight intermediate products

Sections 9.6 and 9.7 have described how post-consumer PET can be reduced into small molecules (i.e., depolymerised) and then re-polymerised to re-create PET or other useful polymers (e.g., PU). Recycling *via* generation of valuable low-MW by-products is not exclusive to post-consumer PET and can be applied to many hydrocarbon-based polymers. Some of these processes can have similarities to the pyrolysis routes (Section 9.14) used to produce fuel products in that they result in the formation of low-MW molecules. The main difference is that usually the pyrolysis process takes place in the absence of oxygen, whereas it can be present in these processes because often oxidation reactions are needed to generate the desired products.

One such recycling approach is described in the patent US6841709, and involves the oxidative decomposition of the polymer by the use of chemicals such as nitrogen dioxide and/or dinitrogen tetraoxide in an inert atmosphere or supercritical carbon dioxide. This method can be applied to any hydrocarbon-based polymer and, in this way, a large range of polar mono- and difunctional chemical compounds can be produced that can be used as feedstocks for the production of other products (or as products in their own right). A second patent, US5516952, partially oxidises polymers using a supercritical (or near supercritical) water mixture. It is claimed that high yields of alkanes, alkenes, aromatics, alcohols, carboxylic acids and ketones, among others, can be produced by this method. The main potential drawbacks to both of these types of processes are their overall complexity and cost.

As mentioned previously, the carefully controlled depolymerisation of waste PET can be used as a means of generating low-MW intermediate products (e.g., monomers and oligomers) that can then be used to create new products, such as crosslinked polyester resins and coatings. However, production of these low-MW intermediates can be looked upon as an end in itself. For example, Achilias and co-workers at the Aristotle University in Greece [41] explored the possibility of depolymerising waste PET from soft drink bottles using an aminolysis process. They carried out the reaction with ethanolamine without a catalyst in a sealed microwave reactor under controlled conditions (i.e., pressure and temperature) and established the activation energy of the reaction using a kinetic model. An analysis of the reaction products by FTIR spectroscopy and DSC showed that the main product present was *bis*(hydroxyethyl) terephthalamide.

Chabert and co-workers [42] claim to have developed a new and original method to recycle PET into low-MW dialkyl-functionalised oligomers. They carried out fast (≈10 min) depolymerisation reactions in a twin-screw laboratory micro-extruder at high temperature with titanium compounds entering into alkoxide ligand exchange reactions with the PET. The structure of the oligomers generated by this system were determined by a combination of ^{1}H-NMR, matrix-assisted laser desorption ionisation time-of-flight and size-exclusion chromatography, and their physical properties characterised by DSC and TGA. The workers proposed ways in which this small-scale process could be scaled up to enable it to be carried out as a reactive extrusion operation using standard equipment.

9.14 Fuel products

One way of recycling waste polymer products (rubbers, thermosets and thermoplastics) is to use them as a source of energy. This route is often described as 'quaternary recycling' and was introduced in Chapter 1. There are several potential ways that this energy can be accessed, with the two principal options being:
- Incineration
- Pyrolysis into fuel products such as oils and gases

The objective of work carried out at Natural State Research Incorporated by Sarker and co-workers [43] was to convert waste PET (which comprised 62.5% carbon, 33.3% oxygen and 4.2% hydrogen) into a liquid hydrocarbon fuel. The fuel that resulted from this research was obtained using a catalyst (calcium hydroxide) at 400–530 °C.

A company called Coskata [44] developed a process to make ethanol from practically any waste source, including household waste, plant waste and waste tyres. The process involves three steps, the first of which thermally depolymerises the feedstock through gasification into its basic building blocks carbon monoxide and hydrogen (called 'syngas'). Syngas is catalytically converted into alcohols, but Coskata feed it to anaerobic bacteria, which convert the carbon monoxide and hydrogen into ethanol. The third step in the process is solvent recovery, in which the ethanol is separated from the fermentation mixture. The process is claimed to have several advantages over traditional corn-based ethanol processes, including improved energy efficiency. It generates 7.7-fold more energy that it consumes and uses ≈1 gallon of water per gallon of ethanol. These figures compare well with those of corn-based ethanol production, which generates 1.3-fold more energy and uses ≈3–4 gallons of water per gallon of ethanol produced.

This recycling option is available to post-consumer PET but not all the systems developed to convert waste plastic into fuel oils can accept the material. For example, Redahan reported in 2011 [45] on the expected opening by 2012 of the first UK facility that could convert end-of-life plastics into road-specification diesel and other kerosene fuels. The facility, which will be opened by SITA in collaboration with the patent holders Cynar, is said to accept plastics such as films, bags and trays that are residual at the end of traditional recycling processes, after the valuable polymer products had been removed from the waste stream, and would usually have to go to landfill or be incinerated. The patented process, which is claimed to produce ≈700 l of full road-specification diesel and 200 l of kerosene for every tonne of plastic waste, can process PE-, PP- and polystyrene (PS)-based waste, but not PET or polyvinyl chloride.

9.15 Miscellaneous products

The use of rPET to manufacture strapping products is mentioned in Chapter 1. WRAP have looked into the possibility of using coloured rPET to produce strapping during the project to investigate the markets for recycling detectable black PET packaging [10]. It was appreciated by the researchers that the standard green strapping (Figure 9.2) could not be made using any significant level of detectable, black

Figure 9.2: Standard green PET strapping. Reproduced with permission from the Waste and Resources Action Programme, Banbury, UK. ©WRAP.

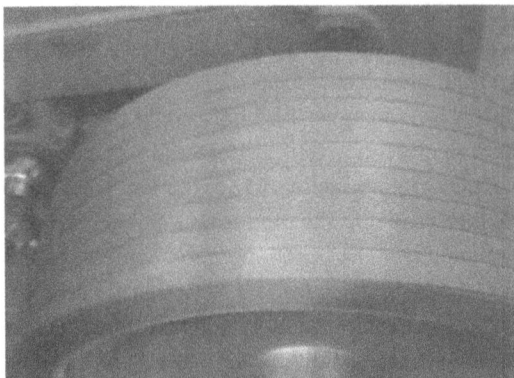

Figure 9.3: Strapping manufactured from detectable black crystalline PET during the WRAP trial. Reproduced with permission from the Waste and Resources Action Programme, Banbury, UK. ©WRAP.

crystalline rPET, and so a trial was run using 100% rPET to evaluate its processing and physical properties. A silvery grey colour strapping product was manufactured without breakage or other operational issues under standard manufacturing conditions (Figure 9.3). It was thought that the silvery grey colour was a result of an increased level of crystallinity that may be induced by the nucleators in the crystalline PET and the stress whitening that results from the orientation and drawing process that strengthens the strapping.

In petroleum wells, the granular, sand-control agents commonly employed are inorganic compounds used to prevent sand from unconsolidated sandstone contaminating the valuable hydrocarbon products. Pereira and Delpech [46] looked into the viability of using rPET as a sand-control agent in an environment that simulated the conditions in an oil well. Pellets of vPET (control) and rPET were confined into metallic cells filled with seawater or petroleum under conditions consistent to those found in sandstone formations in Campos Basin (Rio de Janeiro, Brazil) that are subjected to sand control. The cells were exposed to 70 °C at a pressure of 24.1 MPa for 172 days and were rolled. Samples of the pellets were taken periodically and their thermal, mechanical and granulometric properties evaluated using a range of analytical techniques. Mechanical experiments were done according to API RP 58 and API RP 61 standards to determine the grain-pack permeability by the end of the test period. The particle size of the pellets was also assessed before and after the test. Results showed that neither the vPET nor the rPET pellets underwent significant physical or chemical changes in petroleum or water test media, which suggested that the rPET could be viable as a sand-control agent under these conditions.

Researchers from Montan-Universitat in Leoben, Austria [47] prepared several rPET organo-modified montmorillionite (i.e., organoclay) nanocomposites by melt compounding in a counter-rotating twin-screw extruder. The group then evaluated the topological changes in the polymer matrices relating to the clay modification from dynamic experiments in the shear flow using low-amplitude oscillatory measurements. They observed shear thinning behaviour for all of the organoclay nanocomposites at low frequencies. Some of the organoclays used in the rPET were found to promote degradation reactions, which resulted in reduced values for viscosity and storage modulus at the higher frequencies when compared with the unfilled rPET control sample. By comparison, when the same range of organoclays was compounded into recycled ABS, no such degradation reactions were witnessed in any of the resulting composites.

In addition to the many products covered in this section and Chapter 8, it was reported in 2010 in a general article on the use of environmentally friendly plastic products that rPET was making good progress in being accepted for the manufacturing of general consumer items such as laptop cases and carpets [48]. Another example of the use of rPET for the manufacture of consumer goods was cited in an article in *Plastics and Rubber Asia* in 2009 [49]. This announced that post-consumer PET water bottles had been used as the source of the rPET used to manufacture the housing for the Motorola MotoW233 mobile phone and that this was the first example of this particular re-use application.

Research on the conversion of post-consumer PET and PS into bioplastic polyhydroxyalkanoates has also been carried out [50]. As mentioned in Chapters 2 and 3, the use of bioplastics is increasing at a considerable rate. Their advantages over conventional fossil fuel-derived plastics include their biodegradability and, if produced from plant-derived sources, such as soybean oil, corn oil and fish oil, their sustainability.

The placing into landfill sites of large quantities of post-consumer and industrial PET/PE/PP laminates has been reported by Bansal [51] to be of great concern in India. This has led to research being undertaken by Pluss Polymers and Manas Research and Technology to develop a process which uses additives that can compatibilise this type of waste and convert it into high-value products. Bansal stated that each step of the process was designed to meet the requirements of the different materials that may be present in the laminates. To improve the purity of the waste stream, a complete system for the detection and removal of unwanted material from the industrial laminate waste has also been developed.

9.16 Modification of recycled polyethylene terephthalate to improve properties

rPET can cause problems during processing due to its hydrolytic instability, and this can result in a lowering of MW, which can have a negative impact on the physical properties of the rPET products. This undesirable attribute can cause problems if it is being used to manufacture products described in the sections above. It can also be desirable to improve the compatibility of rPET in composite materials to improve physical properties such as toughness and if the degree of modification is quite considerable. Some examples of the work carried out in this area has been described in Section 9.11. This section contains some recent examples of how researchers have attempted to improve these important properties of rPET.

Srithep and Turng [52] described attempts to improve the process stability of rPET to thermal and hydrolytic degradation by use of a CE and nanoclay. The CE was melt-blended with the rPET in a thermokinetic mixer (K-mixer) and samples prepared using solid and microcellular injection-moulding processes. Testing was carried out on these samples to determine the effects that CE loadings and the simultaneous addition of nanoclays have on their mechanical properties, thermal properties and, in the case of the microcellular samples, their cell morphology. Results showed that addition of 1.3% of the CE enhanced the tensile properties and viscosity of the rPET. If a higher amount (3%) was added, viscosity was increased, but there was less of an improvement in mechanical properties. The solid rPET and CE blends were fairly ductile, but the solid samples containing nanoclay and all the microcellular products showed brittle fracture behaviour. When the cell morphology of the microcellular samples was examined, addition of the nanoclay, and increasing the level of the CE, were found to decrease the average cell size and enlarge cell density.

One of the effects of the instability referred to already is that the MW of the rPET is lower than the original virgin material, and this phenomenon can restrict the products that can be produced from it. This is a problem that has been addressed by several researchers and one such group are from the Fujian Normal University in China [53]. This group evaluated the use of a multiple-functional epoxide CE (ADR4370S) in a melt-mixing environment to increase the MW of rPET. To determine the influence of this extender on the rPET, the effect of adding different levels of the substance on its molecular structure, molecular weight distribution (MWD), chain branching and gel content was investigated by rheology. Results showed that the complex and apparent viscosity of the modified rPET were larger than those of the original rPET. In addition, increments in balancing torque, the reaction peak, and shear thinning behaviour became more pronounced by increasing the concentration of ADR4370S. The reactive modification of the rPET was characterised by the presence of long-chain branching, which in turn resulted in a wider MWD. Chain branching was also correlated to the decrease in the melting point and the crystallisation temperature observed in the modified rPET. A high concentration (1.5%) of ADR4370S resulted in a polymeric structure near the sol–gel transition point whose linear viscoelastic properties obeyed scaling laws, and the relaxation time was found to increase as the amount of ADR4370S increased.

Bimestre and Saron [54] also looked at the possibility of increasing the MW of rPET by the use of CE. The PET waste products from the production of non-woven fabrics were subjected to reactive extrusion in the presence of variable amounts of the secondary stabiliser Irgafos 126. Results showed that Irgafos 126 acted as a CE by increasing the MW of the PET and decreasing its crystallinity. In fact, the changes in the processing properties that resulted were similar to those produced by the well-established CE for PET, pyromellitic dianhydride.

One way of decreasing the problem of a reduction in MW due to the hydrolysis of PET during re-processing is to restrict the amount of water that the pellets absorb. A Japanese group [55] approached this objective by developing 'dry-less pellets' which absorb less moisture and so do not require additional drying before moulding. This result was achieved by a using a new technique, a 'warm air drying system', to dry the strands of PET before they are pelletised, rather than employing the standard method of quenching them in a water bath. The warm air drying system uses warm air to dry the strands of PET slowly on a metal conveyor. The group reported the results of a study carried out to investigate the structural differences within pellets produced using this method with normal pellets produced using the water bath method. Raman microspectrometry, SEM, DSC and Karl Fischer moisture titration were used in this study.

Some researchers have looked at how to improve other properties of rPET, such as its toughness and other physical properties (e.g., tensile strength). For example, a Chinese group [56] used SEBS and a compatibiliser (E-GMA-g-PS) to toughen and compatibilise rPET. The Haake rheocord system was used to characterise the reaction and crystallisation and mechanical properties were investigated to observe the toughening and compatibilisation effect of the SEBS and the compatibiliser. Results showed that the epoxy-functional group of the compatibiliser could react with the end group of the rPET and that the reaction product was very compatible with the SEBS. Data also showed that the crystallisation and nucleation rate were reduced as the addition levels of the SEBS and compatibiliser were increased, and that the notched impact strength could be increased by ≤159%.

Srithep and Turng from Madison University in Wisconsin, US, prepared samples from blends comprising rPET, a CE and nanoclay using solid injection-moulding and microcellular injection-moulding processes [57]. The effects of CE loadings and addition of nanoclays on the thermal and mechanical properties of the sample were recorded. The effects of these two additives on the morphology of the microcellular components were also noted. Addition of CE at 1.3% enhanced the tensile properties and viscosity of the rPET, whereas increasing the level to 3% enhanced the viscosity but resulted in less improvement in tensile properties. The solid rPET and CE blends were found to be fairly ductile, but the rPET samples containing nanoclay, and all the microcellular specimens, showed brittle facture behaviour.

Work carried out by two researchers [58] in India concentrated on the use of fly ash as a filler material for rPET. The pair added fly ash to rPET at several levels at 5–40% *w/w* and determined its effect on the mechanical, rheological, electrical and thermal properties. As part of the same study, they investigated how the particle size of the fly ash affected these properties. The microstructure of the different samples was also investigated using SEM and the observations related to the physical properties. The authors hypothesised that inclusion of microparticles into the matrix resulted in an increase in mechanical properties. The SEM work also revealed that the fly ash was well dispersed within the rPET matrix, and that a very good level of interaction between the fly ash particles and the matrix had taken place.

9.17 Recovery of metals, inorganic and organic compounds from functionalised polyethylene terephthalate products

PET is used to manufacture certain products modified by metals and inorganic compounds (e.g., silver and maghemite) to provide them with the requisite functionality. Examples of these sorts of products are:

– X-ray films
– Magnetic tape
– Prepaid cards

The presence of such substances means that there are limited options available for recycling these products at the end of their lives. A group from Tohoku University [59] in Japan developed a process that could be applied to these types of PET products to reclaim the functional additives and, ultimately, obtain benzene-rich oil from the base polymer. The process involved hydrolysing the PET using steam at 450 °C and the hydrolysis products included a metal-containing carbonaceous residue and TPA. The group demonstrated that the metals could be recovered quantitatively and silver and inorganic compounds, such as maghemite and anatase, were recovered without changes in their crystal structures or compositions. In a second step, the TPA was decarboxylised in the presence of calcium oxide at 700 °C to produce benzene with an average yield of 34% and purity of 76%.

References

1. E. Rahmani, M. Dehestani, M.H.A Beygi, H. Allahyari and I.M. Nikbin, *Construction and Building Materials*, 2013, **47**, 1, 1302.
2. Anon, *PETplanet Insider*, 2010, **11**, 9–10, 22.
3. Anon, *PETplanet Insider*, 2012, **13**, 10, 22.
4. A. Stoilkov and R. Knief in *Proceedings of the Polyurethanes 2010 Technical Conference*, Houston, TX, USA, 11–13th September 2010, Eds., American Chemistry Council, Centre for the Polyurethanes Industry, Arlington, VA, USA, 2010, Paper No.48, p.11.
5. P. Malnati, *Composites Technology*, 2011, **17**, 2, 46.
6. R. Stewart, *Reinforced Plastics*, 2010, **54**, 4, 321.
7. Anon, *International Fiber Journal*, 2012, **26**, 1, 6.

8. H.J. Koo, G.S. Chang, S.H. Kim, W.G. Hahm and S.Y. Park, *Fibres and Polymers*, 2013, **14**, 12, 2083.
9. G. Kostov, A. Atanassov and D. Kiryakova, *Progress in Rubber, Plastics and Recycling Technology*, 2013, **29**, 4, 255.
10. *End-Markets for Recycled Detectable Black PET Plastics*, Final Report, Waste and Resources Action Programme (WRAP), Banbury, UK, July 2013.
11. Anon, *PETplanet Insider*, 2013, **14**, 4, 29.
12. L. Di, *International Fiber Journal*, 2012, **26**, 1, 28.
13. P.S. Upasani, A.K. Jain, N. Save, U.S. Agarwal and A.K. Kelkar, *Journal of Applied Polymer Science*, 2012, **123**, 1, 520.
14. L.L. Hayes, *AATCC Review*, 2011, **11**, 4, 37.
15. S. Levy, *Nonwovens Industry*, 2011, **42**, 3, 52.
16. M.Y. Abdelaal, T.R. Sobahi and M.S.I. Makki, *Construction and Building Materials*, 2011, **25**, 8, 3267.
17. A.R. Zahedi, M. Rafizadeh and S.R. Ghafarian, *Polymer International*, 2009, **58**, 9, 1084.
18. A.B. Juraev, R.I. Adilov, T.A. Nizamov, M.G. Alimukhamedov, F.A. Magrupov and I.T. Usmonov, *Kautschuk Gummi Kunststoffe*, 2014, **67**, 3, 41.
19. S. Katoch, V. Sharma and P.P. Kundu, *Chemical Engineering Science*, 2010, **65**, 15, 4378.
20. Rama Konduri and R.W. Fonseca in *Proceedings of the 69th SPE ANTEC Conference*, Boston, MA, USA, 1–5th May, Ed., Society of Plastics Engineers, Brookfield, USA, 2011, p.2660.
21. A.M. Issam, S. Hena and A.K. Nurul Khizrien, *Journal of Polymers and the Environment*, 2012, **20**, 2, 469.
22. D.S. Dias, M.S. Crespi, C.A. Ribeiro and M. Kobelnik, *Journal of Applied Polymer Science*, 2011, **119**, 3, 1316.
23. M. Kathalewar, N. Dhopatkar, B. Pacharane, A. Sabnis, P. Raut and V. Bhave, *Progress in Organic Coatings*, 2013, **76**, 1, 147.
24. Anon, *British Plastics and Rubber*, 2013, April, 37.
25. J. Kajaks, K. Kalnins, S. Reihmane and A. Bernava, *Progress in Rubber, Plastics and Recycling Technology*, 2014, **30**, 2, 87.
26. Anon, *High Performance Plastics*, 2014, January, 4.
27. V. Yatish, K.D. Patel, D.P. Sule and D. Devanshu in *Proceedings of the ANTEC 2012 Conference*, Mumbai, India, 6–7th December, Ed., Society of Plastics Engineers, Brookfield, CT, USA, 2012, p.927.
28. N.N.B. Mohammad and A. Arsad, *Malaysian Polymer Journal*, 2013, **8**, 1, 8.
29. S. Kayaisang, S. Saikrasun and T. Amornsakchai, *Journal of Polymers and the Environment*, 2013, **21**, 1, 191.
30. S. Sombatdee, S. Saikrasun and T. Amornsakchai, *Journal of Reinforced Plastics and Composites*, 2009, **28**, 24, 2983.
31. N. Mondadori, R. Nunes, L. Canto and A. Zattera, *Journal of Thermoplastic Composite Materials*, 2012, **25**, 6, 747.
32. S. Thumsorn, K. Yamada, Y.W. Leong and H. Hamada in *Proceedings of the 69th SPE ANTEC Conference*, Boston, MA, USA, 1–5th May, Ed., Society of Plastics Engineers, Brookfield, USA, 2011, p.2231.
33. Y. Srithep, A. Javadi, S. Pilla, L-S. Turng, S. Gong, C. Clemons and J. Peng, *Polymer Engineering and Science*, 2011, **51**, 6, 1023.
34. R. van der Meer and V. Frenz in *Proceedings of the 6th European Conference on 'Additives and Colours'*, Antwerp, Belgium, 11-12th March, Eds., Ed., Society of Plastics Engineers, Additives & Color Europe Division, Antwerp, Belgium, 2009, Paper No.18, p.3.

35. N. Kunimune, K. Yamada, Yew Wei Leong, S. Thumsorn and H. Hamada, *Journal of Applied Polymer Science*, 2011, **120**, 1, 50.
36. I. Kelnar, V. Sukhanov, J. Rotrekl and L. Kapralkova, *Journal of Applied Polymer Science*, 2010, **116**, 6, 3621.
37. R.N. Baxi, S.U. Pathak and D.R. Peshwe, *Journal of Applied Polymer Science*, 2010, **115**, 2, 928.
38. H. Cornier-Rios, P.A. Sundarum and J.T. Celorie, *Journal of Polymers and the Environment*, 2007, **15**, 1, 51.
39. C.T. Ferreira, J.B. Da Fonseca and C. Saron, *Polimeros: Ciencia e Tecnologia*, 2011, **21**, 2, 118.
40. H. Nabil and H. Ismail, *Progress in Rubber, Plastics and Recycling Technology*, 2014, **30**, 2, 115.
41. D.S. Achilias, G.P. Tsintzou, A.K. Nikolaidis, D.N. Bikiaris and G.P. Karayannidis, *Polymer International*, 2011, **60**, 3, 500.
42. M. Chabert, V. Bounor-Legare, N. Mignard, P. Cassagnau, C. Chamignon and F. Boisson, *Polymer Degradation and Stability*, 2014, **102**, 1, 122.
43. M. Sarker, A. Kabir, M.M. Rashid, M. Molla and A.S.M. Mohammad Din, *Journal of Fundamentals of Renewable Energy and Applications*, 2011, Paper R101202, 5.
44. M. Bryner, *Chemical Week*, 2008, **170**, 10, 53.
45. E. Redahan, *Materials World*, 2011, **19**, 10, 3.
46. A.Z.I Pereira and M.C. Delpech, *Polymer Degradation and Stability*, 2012, **97**, 7, 1158.
47. M. Kracalik, S. Laske, A. Witschnigg and C. Holzer, *Macromolecular Symposia*, 2012, **311**, 33.
48. Anon, *Plastics and Rubber Asia*, 2010, **25**, 177, 8.
49. Anon, *Plastics and Rubber Asia*, 2009, **24**, 163, 8.
50. J. Evans, *Plastics Engineering*, 2010, **66**, 2, 14.
51. N. Bansal, *Popular Plastics and Packaging*, 2011, **56**, 4, 49.
52. Y. Srithep and L. Turng, *Journal of Polymer Engineering*, 2014, **34**, 1, 5.
53. X. Liren, W. Hai, Q. Qingrong, J. Xia, L. Xinping, H. Baoquan and C. Qinghua, *Polymer Engineering and Science*, 2012, **52**, 10, 2127.
54. B.H. Bimestre and C. Saron, *Materials Research*, 2012, **15**, 3, 467.
55. M. Setomoto, K. Iyagawa, H. Inoya, K. Yamada and H. Hamada in *Proceedings of the 70th SPE ANTEC Conference*, Orlando, FL, USA, 2–4th April 2012, Ed., Society of Plastics Engineers, Brookfield, CT, USA, 2012, p.4.
56. Q-Q.Tang, B. Yang and C-X. Zhou, *Journal of Functional Polymers*, 2005, **18**, 3, 504.
57. Y. Srithep and L-S. Turng in *Proceedings of the 69th SPE ANTEC Conference*, Boston, MA, USA, 1–5th May, Ed., Society of Plastics Engineers, Brookfield, USA, 2011, p.2145.
58. A.K. Sharma and P.A. Mahanwar, *International Journal of Plastics Technology*, 2010, **14**, 1, 53.
59. S. Kumagai, G. Grause, T. Kameda and T. Yoshioka, *Environmental Science and Technology*, 2014, **48**, 6, 3430.

10 Conclusion

The recycling of polyethylene terephthalate (PET) has been carried out for over 40 years, but this book has shown how increasing environmental, economic and societal pressures in recent years has resulted in an exponential increase in interest being shown in the development of effective and commercially viable methods for recycling this material. These pressures, coupled with the research and development initiatives of industry, and assisted by funding by government, has enabled several recycling processes and systems to be developed and optimised for PET and many other plastics. These have ranged from new sorting and separation systems, to 'super-clean' recycling processes to enable the production of high-quality, food-grade recycled polyethylene terephthalate (rPET) capable of being used at the 100% level to manufacture new food contact materials and articles.

The studies that have been carried out in academia and within industry have demonstrated that rPET can be re-used very successfully in the manufacture of a wide range of products. In addition to its use for the manufacture of new PET products for food packaging (e.g., bottles and trays) and non-food-grade products (e.g., fibres, clothing and strapping), this book has shown that there are many other possible routes available for the recycling of waste PET and post-consumer PET: conversion into fuel products; blending with rubbers and plastics to create new materials capable of being used in a range of products; its use to produce construction and automotive products; depolymerisation to create monomers and other low-molecular weight intermediate molecules that can be used to manufacture new polymers (e.g., polyurethane coatings) and thermosetting materials (e.g., unsaturated polyester resins).

As this book shows, after many years of research, development and characterisation work, the recycling of PET is a vibrant, growing industry that should continue to gain market share over the next few years. Great progress has been made, but the use of rPET in food packaging is an area that is still generating a lot of interest and in which a considerable amount of research is being carried out. Also, there is appreciable research activity in areas in which there are problems to be solved for the rPET industry and other plastic recycling sectors. One prime example is the sorting of highly coloured and black items in the plastic waste stream, a lot of which is food-grade PET packaging. This problem is being tackled by the development of new detection systems and creation of new black pigments that do not interfere with conventional near-infrared spectroscopy detection systems. Also, an area that presents large challenges (but which would enable a large amount of valuable material to be accessed for recycling if a solution was found) is the separation and recycling of complex laminates and multiple-layer films containing PET and other polymers.

Regulatory systems are in place in many parts of the world and the technical argument has essentially been won for the effective use of post-consumer PET in

https://doi.org/10.1515/9783110640304-010

many products. However, one of the principal factors restricting commercial exploitation is the cost of the rPET relative to that of the virgin PET product. With the decrease in fossil fuel prices in recent years, and the downward pressure expected to exert on the cost of virgin plastics, this situation could get worse in the short term. However, it is to be hoped that other factors, such as stricter governmental targets for recycling, environmental costs (e.g., landfill charges), and better long-term planning will assist the commercial viability and successful exploitation of rPET in the future. These factors, together with continual improvements in the collection and sorting of post-consumer PET, continued growth of 'green' corporate strategies, and the growth opportunities available for PET in developing countries, should ensure that the global market for rPET will continue to increase for many years to come.

Abbreviations

ABS	Acrylonitrile–butadiene–styrene
AC	Acetaldehyde
APET	Amorphous polyethylene terephthalate
APR	US Association of Postconsumer Plastic Recyclers
ASE	Accelerated solvent extraction
ASTM	American Society for Testing and Materials
B2B	Bottle-to-bottle
BHET	*Bis*(2-hydroxyethyl)terephthalate
CAGR	Compound annual growth rate
CARE	Consortium for Automotive Recycling
CE	Chain extender(s)
CEN	European Committee for Standardization
CFR	Code of Federal Regulations
CPET	Crystalline polyethylene terephthalate
CSD	Carbonated soft drinks
CTAB	Cetyltrimethylammonium bromide
DEG	Diethylene glycol
DSC	Differential scanning calorimetry
EC	European Commission
EFSA	European Food Safety Authority
EG	Ethylene glycol
E-GMA	Ethylene-glycidyl methacylate copolymer
EBM	Extrusion blow moulding
EPA	Environmental Protection Agency
EU	European Union
FAC	Fly ash cenosphere(s)
FCC	Food contact control
FDA	US Food and Drug Administration
FID	Flame ionisation detector
FTIR	Fourier-Transform infrared spectroscopy
FTR	Flake-to-resin
FWO	Flynn–Wall–Ozawa method
GC	Gas chromatography
GMP	Good manufacturing practice
GPC	Gel permeation chromatography
GRAS	Generally regarded as safe
HDPE	High-density polyethylene
HNT	Halloysite nanotubes
HQWF	High-quality washed flake
ICP-AES	Inductively coupled plasma-atomic emission spectroscopy
ILSI	International Life Sciences Institute
IPP	Integrated Product Policy
IPPC	Integrated Pollution Prevention and Control
IR	Infrared
ISO	International Organization for Standardization
IT	4-Isopropyltoluene

https://doi.org/10.1515/9783110640304-011

LCA	Lifecycle assessment
LCP	Liquid crystal polymer
LDPE	Low-density polyethylene
LIBS	Laser-induced breakdown spectroscopy
Lim	Limonene
LIPS	Laser-induced plasma spectroscopy
MA	Maleic anhydride
MAF	Mobile amorphous fraction
MALDI	Matrix-assisted laser desorption ionisation
MD	2-Methyl-1,3-dioxolane
MEG	Monoethylene glycol
MIR	Mid-infrared spectroscopy
MRF	Materials recycling facility(ies)
MRS	Multiple rotation system
MS	Mass spectrometry
MW	Molecular weight
MWD	Molecular weight distribution
NAPCOR	US National Association for PET Container Resources
NIR	Near-infrared spectroscopy
NMR	Nuclear magnetic resonance spectroscopy
NPG	Neopentyl glycol
NR	Natural rubber
PA	Polyamide(s)
PAT	Process analytical technology
PBAT	Poly(butylene adipate-*co*-terephthalate)
PBT	Polybutylene terephthalate
PE	Polyethylene
PEF	Polyethylene furanoate
PEG	Polyethylene glycol
PET	Polyethylene terephthalate
PG	Propylene glycol
PLA	Polylactic acid
PP	Polypropylene
PRE	Plastics Recyclers Europe
PRF	Plastic-recycling facility
PS	Polystyrene
PU	Polyurethane
PUF	Polyurethane foam(s)
PVC	Polyvinyl chloride
Py	Pyrolysis
R&D	Research and development
RAF	Rigid amorphous fraction
RCRA	Resource Conservation and Recovery Act
RECOUP	Recycling Of Used Plastics Limited
RF	Radio-frequency
RFID	Radio-frequency identification tags
rPET	Recycled polyethylene terephthalate
rPP	Recycled polypropylene
RSM	Response surface method

SEBS	Styrene–ethylene–butylene–styrene
SEM	Scanning electron microscopy
SML	Specific migration limit
SPI	Society of the Plastics Industry
SSP	Solid-state polymerisation
SSSP	Solid-state shear pulverisation
T_c	Crystallisation peak temperature
TEG	Triethylene glycol
TGA	Thermogravimetric analysis
T_m	Melting temperature
TPA	Terephthalic acid
TPE	Thermoplastic elastomer(s)
UV	Ultraviolet
vPET	Virgin polyethylene terephthalate
WRAP	Waste and Resources Action Programme
XRD	X-ray diffraction
XRF	X-ray fluorescence spectroscopy

Index

https://doi.org/10.1515/9783110640304-012

www.ingramcontent.com/pod-product-compliance
Lightning Source LLC
Chambersburg PA
CBHW061415210326
41598CB00035B/6225